Human-AI Empowerment

As artificial intelligence (AI) advances, the question of how AI can empower humans over the long term has become increasingly important. This book, *Human-AI Empowerment* (HAIE), provides a timely exploration of strategies for aligning AI with long-term human goals, ensuring that AI acts as an empowering force across multiple dimensions. Drawing on interdisciplinary research from fields such as AI, HCI, psychology, education, economics, and social science, the book develops comprehensive frameworks for studying and optimizing AI's impact on human empowerment. HAIE investigates empowerment from a human-centered computing (HCC) perspective, examining how AI systems can track and adapt to progressively achieve long-term goals. The book explores techniques for fostering a mutually beneficial human-AI synergy, delving into AI Empowerment approaches, applicable Human-Computer Interaction methods for long-term engagement, and insights from various disciplines on long-term goal management. Through integrative frameworks, empirical evidence, and ongoing work in the field, this volume informs academics and practitioners seeking to harness AI as a transformative technology for concretely empowering humanity. This book highlights the need for comprehensive approaches to understanding and shaping the future of human-AI collaboration, maximizing its potential to expand human possibilities and support the pursuit of mid-term and long-term goals.

Human-AI Empowerment

An Interdisciplinary Perspective

Carlos Toxtli-Hernández

CRC Press
Taylor & Francis Group
Boca Raton London New York

CRC Press is an imprint of the
Taylor & Francis Group, an **informa** business
A CHAPMAN & HALL BOOK

First edition published 2026
by CRC Press
2385 NW Executive Center Drive, Suite 320, Boca Raton FL 33431

and by CRC Press
4 Park Square, Milton Park, Abingdon, Oxon, OX14 4RN

CRC Press is an imprint of Taylor & Francis Group, LLC

© 2026 Taylor & Francis Group, LLC

ISBN: 978-1-032-88202-4 (hbk)
ISBN: 978-1-032-88204-8 (pbk)
ISBN: 978-1-003-53662-8 (ebk)

DOI 10.1201/9781003536628

Typeset in Latin Modern font
by KnowledgeWorks Global Ltd.

Contents

List of Figures

List of Abbreviations

AI artificial intelligence

ANOVA analysis of variance

GIS geographic information system

HCC human-centered computing

HCI human-computer interaction

ICT information and communication technology

IoT Internet of Things

RS remote sensing

Preface

The integration of artificial intelligence (AI) into the fabric of human existence is no longer a distant prospect; it is a rapidly unfolding reality. AI systems increasingly mediate our communications, shape our consumption, influence our decisions, and augment our professional practices. Amidst the undeniable capabilities these technologies offer, a critical question emerges with growing urgency: How do we ensure that AI serves not merely as an instrument of efficiency or automation, but as a genuine catalyst for human empowerment? This question transcends purely technical considerations, touching upon fundamental aspects of human agency, societal values, and the long-term trajectory of human development.

The challenge of aligning AI with human interests is often framed in terms of preventing catastrophic risks or ensuring ethical compliance. While crucial, this perspective can sometimes overshadow the proactive and equally vital goal of designing AI to actively enhance human potential and expand the "capability set" – the real freedoms individuals have to achieve outcomes they value. "Human-AI Empowerment," the book you hold, is dedicated to exploring this proactive agenda. It delves into the deliberate construction of AI systems and interaction paradigms that foster, rather than diminish, human autonomy, competence, and the capacity for meaningful action in the pursuit of self-defined, long-term goals.

This endeavor inherently necessitates bridging diverse intellectual traditions. It requires moving beyond siloed approaches where AI development is solely the domain of computer scientists, or user interaction solely the concern of Human-Computer Interaction (HCI) practitioners.

Understanding and fostering empowerment demands a synthesis. We must draw upon the analytical rigor of AI and machine learning; the user-centric design methodologies of HCI; the nuanced models of motivation, learning, and cognition from psychology; the pedagogical insights from education science; the frameworks for analyzing opportunity and well-being from economics and development studies; and the critical perspectives on technology's societal role from sociology and philosophy of technology. This book brings these interdisciplinary perspectives into dialogue, seeking a more holistic understanding of the conditions under which AI can truly empower.

The structure of this volume reflects this multifaceted approach. **Part I** establishes the conceptual groundwork, defining Human-AI Empowerment and exploring the key theoretical frameworks and ethical considerations that underpin this field. **Part II** focuses on methodology, presenting frameworks specifically designed for conducting longitudinal studies – essential for assessing the often subtle, cumulative effects of AI interaction on human capabilities and opportunities over time. **Part III** delves into actionable strategies, examining AI-driven approaches to empowerment, HCI techniques for sustaining long-term engagement, and cross-disciplinary insights into supporting individuals' management of their long-term goals. **Part IV** provides empirical grounding through diverse case studies, showcasing concrete examples where AI technologies are demonstrably boosting human potential across domains like healthcare, education, and the creative industries. Finally, **Part V** looks ahead, discussing ongoing work, persistent challenges (such as scalability and equity), the crucial role of policy and governance, and future research directions in the dynamic field of Human-AI Empowerment.

This book is intended for a broad audience grappling with the implications of AI: researchers seeking robust methodologies, practitioners designing next-generation AI systems, policymakers shaping the regulatory landscape, and students preparing to navigate and contribute to an AI-infused world. We aim to provide not just theoretical discussion, but

actionable insights and frameworks for creating AI that genuinely enhances human agency and aligns with our deepest
values and aspirations. The ultimate goal is to foster a more
critical, informed, and proactive approach to AI development
– one that moves beyond passive acceptance toward the intentional shaping of technology to serve enduring human goals.

The creation of this work has been a journey enriched by
collaboration and dialogue with experts across many fields. I
extend my sincere gratitude to all contributors whose insights
have shaped this volume. My thanks also go to the editorial
team at Taylor & Francis for their invaluable support throughout the publication process.

As AI's capabilities continue to expand, the need for critical reflection and deliberate design choices becomes ever more
paramount. This book is an invitation to that reflection, a contribution to the ongoing conversation about the kind of future
we wish to build with artificial intelligence. It is my hope that
the concepts and frameworks presented here will stimulate further research, inspire innovative designs, and contribute to the
realization of a future where human and artificial intelligence
collaborate synergistically to unlock the full potential of both.

Carlos Toxtli-Hernández
Clemson University
2025

Acknowledgments

I would like to extend my heartfelt gratitude to the many individuals who have contributed to the creation of this book. First and foremost, I would like to thank my colleagues and collaborators from various institutions and disciplines, whose insights and expertise have been invaluable in shaping the content of this book. Their dedication to advancing the field of Human-Centered AI has been a constant source of inspiration.

I am also deeply grateful to the researchers and practitioners who have generously shared their knowledge and experiences through case studies, interviews, and discussions. Their contributions have enriched the book with real-world examples and practical insights, bringing the concepts and frameworks to life.

Special thanks go to the editorial team at Taylor & Francis for their unwavering support, guidance, and patience throughout the publishing process. Their expertise and professionalism have been instrumental in bringing this book to fruition.

Finally, I would like to express my profound appreciation to my family and friends for their unconditional support, encouragement, and understanding throughout this journey. Their belief in the importance of this work has been a driving force behind my efforts to contribute to the responsible development and deployment of AI technologies for human empowerment.

In line with Taylor & Francis' policy on the use of artificial intelligence tools, the author acknowledges the use of specific AI assistance during the preparation of this manuscript. Tools such as scite.AI were employed to aid in the identification of potentially relevant citations during the literature review process. Furthermore, Large Language Models including Claude AI (Anthropic) and ChatGPT (OpenAI), along with the tool Grammarly, were utilized solely for the purposes of improving language, grammar, style, and clarity of the text. Crucially, these tools were not used for generating the core concepts, arguments, research content, data analysis, or overall structure of the manuscript. All substantive ideas, analyses, theoretical frameworks, and conclusions presented are the original work of the author, developed prior to and independent of the use of these language enhancement tools. All outputs generated by these AI tools were carefully reviewed, critically evaluated, edited, and verified by the author to ensure accuracy, alignment with the intended meaning, and adherence to scholarly standards. The author remains fully accountable for the originality, validity, and integrity of the entire content presented in this book.

About the Author

Carlos Toxtli-Hernández is an Assistant Professor at Clemson University, where he focuses on the study of Human-Centered Artificial Intelligence. His research explores the intersection of artificial intelligence, human-computer interaction, and automation, with a particular emphasis on developing methodologies and frameworks for studying the long-term impact of AI on human empowerment. He has published extensively in leading academic journals and conferences, and his work has been recognized with numerous awards and grants. He holds a Ph.D. degree in Computer Science from Northeastern University.

Introduction

The contemporary landscape is undeniably shaped by the accelerating advancements in artificial intelligence (AI). From automating complex analyses to generating creative content, AI technologies permeate diverse facets of human activity, promising unprecedented gains in efficiency and capability. This technological surge, however, brings forth a critical juncture: alongside the potential for profound societal benefit lies the risk of unintended consequences, including the potential erosion of human agency, the exacerbation of inequalities, and misalignment with long-term human flourishing [280, 44]. The central challenge, therefore, is not merely to build more powerful AI, but to ensure its development and deployment actively *empower* humanity [364].

This book confronts this challenge directly. It proposes and explores a specific paradigm: the deliberate design and evaluation of AI systems aimed at enhancing human capabilities, expanding opportunities, and facilitating the pursuit and achievement of meaningful, self-defined goals over the long term. This perspective moves beyond viewing AI simply as a tool for task automation or immediate problem-solving. Instead, it positions AI as a potential catalyst for human development, fostering growth in skills, knowledge, and the fundamental freedoms individuals possess to live lives they value [292, 235]. Achieving this necessitates a deep understanding of the intricate interplay between technological design, human psychology, socio-economic structures, and ethical considerations [106].

Our approach is firmly rooted in Human-Centered Computing (HCC), a perspective emphasizing that technology should amplify human potential through a focus on usability, agency, and control [298]. While traditional HCI often centers on immediate usability and task efficiency, an HCC lens applied to AI empowerment compels us to consider the *longitudinal* effects of AI interaction. How do these systems shape human skills, self-perception, motivation, and goal-setting behaviors over months or years? This requires designing AI not just to be *used*, but to be a partner in growth, respecting user autonomy while providing scaffolding for development [234]. This human-centric stance is crucial for navigating the potential pitfalls of AI, such as algorithmic bias, deskilling, or the creation of dependency, ensuring that technology serves human ends [232].

Addressing the complexities of Human-AI Empowerment inherently demands an interdisciplinary synthesis. Computer science and AI provide the technical foundations, yet they are insufficient alone. Human-Computer Interaction (HCI) offers methods for designing intuitive and engaging interfaces for long-term use. Psychology furnishes critical models of motivation, learning, goal pursuit, and cognitive augmentation (e.g., Self-Determination Theory [281]). Education research informs the design of AI-driven learning and skill development systems. Economics provides frameworks, like the Capability Approach [293, 13], for conceptualizing and measuring expanded opportunities. Sociology and social science offer lenses to analyze the societal adoption patterns, equity implications, and potential systemic impacts of widespread AI deployment [369]. Philosophy provides the ethical grounding necessary to navigate value alignment and the normative dimensions of empowerment. This book endeavors to weave these diverse threads into a coherent understanding.

Consequently, *Human-AI Empowerment* pursues three primary objectives:

1. **Develop Methodological Frameworks:** To move beyond anecdotal evidence, we aim to synthesize interdisciplinary research to construct rigorous methodological

frameworks for empirically assessing AI's longitudinal impact on human capabilities, opportunities, and potential. This involves defining measurable constructs of empowerment and outlining quantitative, qualitative, and mixed-methods approaches suitable for long-term study.

2. **Analyze Power Dynamics and Disempowerment Risks:** We critically examine the complex dynamics of power inherent in human-AI relationships. This includes identifying and analyzing potential mechanisms of disempowerment—such as algorithmic bias, cognitive deskilling, attention manipulation, and the creation of dependencies—and proposing concrete strategies for their mitigation.

3. **Articulate Design Principles for Empowering AI:** We seek to translate theoretical insights and empirical findings into actionable design principles and interaction paradigms. The goal is to guide the creation of AI systems that foster sustained engagement, actively promote skill development and critical thinking, adapt to users' evolving long-term goals, and demonstrably align with human values.

Through addressing these objectives, this book aims to provide researchers, designers, developers, policymakers, and students with a foundational understanding and practical tools for harnessing AI as a transformative force for genuine human empowerment. It seeks to contribute to a future where AI and humanity collaborate synergistically, unlocking potential and fostering progress toward a more capable, equitable, and fulfilling world.

I

Foundations of Human-AI Empowerment

The pursuit of artificial intelligence (AI) that genuinely enhances the human condition requires moving beyond metrics of task efficiency or computational power. It demands a foundational shift toward understanding how AI can become a partner in human growth and goal achievement. This part lays the groundwork for such an understanding, establishing the core tenets of Human-AI Empowerment. We begin by carefully defining this paradigm, distinguishing it from related concepts and exploring its multidimensional nature. We then delve into the key theoretical frameworks drawn from diverse disciplines that provide the intellectual scaffolding for this approach. Subsequently, we trace the historical evolution of human-centered thinking within AI and computing, contextualizing the emergence of empowerment as a specific focus. Finally, we confront the critical ethical considerations inherent in designing and deploying systems intended to shape human potential, recognizing that ethical alignment is not an add-on but a fundamental prerequisite for true empowerment.

DOI: 10.1201/9781003536628-1

1.1   DEFINING HUMAN-AI EMPOWERMENT

Human-AI Empowerment signifies a paradigm shift in the design, development, and evaluation of AI. It moves beyond viewing AI merely as a tool for automation or assistance toward conceptualizing it as a catalyst for long-term human development. Specifically, Human-AI Empowerment refers to the intentional creation and application of AI systems designed to measurably enhance human capabilities, expand the set of real opportunities available to individuals and communities, and facilitate the pursuit and achievement of their self-defined, long-term goals. This paradigm emphasizes a symbiotic, co-adaptive relationship where both human potential and AI utility grow over time through sustained interaction [194].

At its heart, this definition centers on human agency and potential. It draws inspiration from psychological theories like Self-Determination Theory, which posits that fostering autonomy, competence, and relatedness is crucial for intrinsic motivation and well-being [281]. Empowering AI, therefore, should not dictate goals but rather provide resources, insights, and scaffolding that support users in identifying and pursuing their own aspirations. It must enhance competence not just in using the AI tool itself, but in the underlying domain or skill the tool mediates (e.g., enhancing writing skills, not just producing text faster). It might even foster relatedness through facilitating meaningful connections or collaborations mediated by AI [84].

Furthermore, the definition aligns with the Capability Approach prominent in development economics and philosophy [293, 235]. From this perspective, empowerment is not solely about providing resources (like AI tools), but about expanding individuals' capabilities—their effective freedom to achieve valued "functionings" (states of being and doing, such as being knowledgeable, healthy, or socially engaged) [13]. Thus, empowering AI must be evaluated not just on task performance metrics, but on its contribution to expanding users' repertoire of achievable goals and life paths. Does the AI tool enable someone to pursue a new career, learn a complex skill they previously could not, or participate more fully in civic life? This necessitates a deliberate focus on the longitudinal dimension. Unlike systems optimized for short-term task completion, empowering AI considers the cumulative impact of interaction. It involves designing for sustained engagement, progressive skill development, and adaptation to users' evolving needs and goals over months or years [169]. This long-term perspective inherently recognizes that both the human user and the AI system may change and adapt through their interaction, leading to a co-evolutionary dynamic [145].

Technologically, this implies designing AI systems that are not only capable but also interpretable, adaptable, and aligned with human cognitive processes. They should function as augmentative tools, leveraging AI's strengths in data processing and pattern recognition while respecting and enhancing human strengths like creativity, critical thinking, and ethical judgment [91, 298]. As visualized in Figure 1.1, this involves integrating human needs and AI capabilities within specific contexts to foster augmented intelligence and adaptive support, leading to outcomes like enhanced capabilities, expanded opportunities, goal achievement, and overall well-being, all guided by ethical alignment and a long-term perspective.

Crucially, this definition acknowledges the inherent power dynamics in human-AI interaction [310]. Empowerment requires mitigating risks of deskilling, dependence, manipulation, or algorithmic bias that could undermine agency or exacerbate inequalities [90]. Therefore, ethical considerations, transparency, and user control are not peripheral features but integral components of the definition itself. True Human-AI Empowerment aims for a future where AI serves as a responsible partner in unlocking human potential across diverse domains and populations.

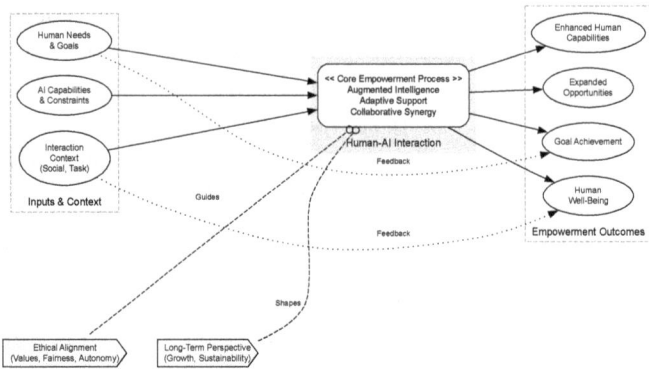

Figure 1.1   A Framework for Human-AI Empowerment

1.2 KEY CONCEPTS AND THEORETICAL FRAMEWORKS

The field of Human-AI Empowerment is underpinned by a rich tapestry of concepts and theoretical frameworks drawn from diverse disciplines such as computer science, psychology, sociology, and philosophy. These foundational elements provide the intellectual scaffolding necessary for understanding the complex interactions between humans and AI systems, and for designing interventions that genuinely enhance human capabilities and opportunities. This section explores the key concepts and theoretical frameworks that inform the study and practice of Human-AI Empowerment, elucidating their relevance and applications in this emerging field. Figure 1.2 provides a visual overview of these diverse disciplinary contributions.

One of the central concepts in Human-AI Empowerment is that of augmented intelligence, which posits that AI systems should be designed to enhance and complement human cognitive abilities rather than to replace them. This concept builds on the work of researchers like Douglas Engelbart, who envisioned technology as a means to augment human intellect and problem-solving capabilities [91]. In the context of Human-AI

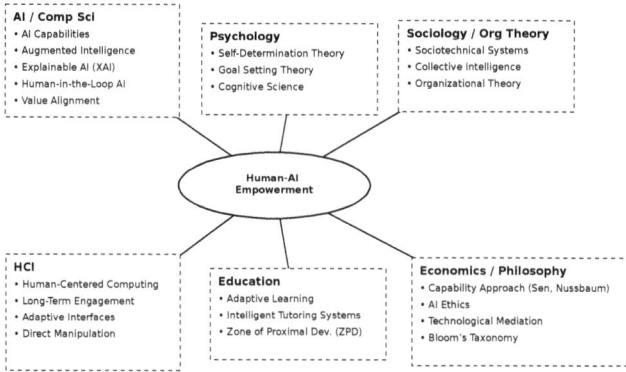

Figure 1.2   Interdisciplinary Roots of Human-AI Empowerment

Empowerment, augmented intelligence manifests in AI systems that provide contextual information, suggest alternative perspectives, or automate routine cognitive tasks, thereby freeing human cognitive resources for higher-order thinking and creativity. The theoretical underpinnings of augmented intelligence draw from cognitive science and human factors engineering, emphasizing the importance of understanding human cognitive processes and limitations in designing effective AI-assisted tools and interfaces.

Another key concept is that of adaptive learning systems, which forms a crucial component of long-term Human-AI Empowerment. Adaptive learning systems are AI-powered educational tools that dynamically adjust their content, pace, and pedagogical approach based on the learner's progress, preferences, and cognitive state. This concept is grounded in educational psychology theories such as Vygotsky's Zone of Proximal Development [346] and Bloom's Taxonomy of Educational Objectives [39]. In the context of Human-AI Empowerment, adaptive learning systems extend beyond traditional educational settings, encompassing lifelong learning and skill development in various domains. These systems leverage machine learning algorithms to create personalized learning experiences that optimize knowledge acquisition and retention, fostering continuous growth and empowerment over extended periods. The concept of human-in-the-loop AI is also fundamental to Human-AI Empowerment, emphasizing the importance of maintaining human oversight and input in AI-driven processes. This approach recognizes that while AI systems can process vast amounts of data and identify patterns beyond human capabilities, human judgment remains crucial for context-sensitive decision-making, ethical considerations, and creative problem-solving. The theoretical basis for human-in-the-loop AI draws from cybernetics and control theory, as well as more recent work in interactive machine learning [96]. In Human-AI Empowerment scenarios, this concept translates into the design of AI systems that actively seek human input at critical junctures, provide explanations for their recommendations, and allow for easy human intervention and correction.

The theoretical framework of distributed cognition provides another crucial lens through which to understand Human-AI Empowerment. Developed by cognitive scientists like Edwin Hutchins [148], distributed cognition posits that cognitive processes extend beyond the individual mind to encompass interactions with external tools, environments, and other individuals. In the context of Human-AI Empowerment, this framework helps conceptualize the human-AI relationship as a cognitive ecosystem in which knowledge, memory, and problem-solving capabilities are distributed between human users and AI systems. This perspective informs the design of AI interfaces and interaction modalities that seamlessly integrate with human cognitive processes, creating a symbiotic relationship that enhances overall cognitive performance.

The capability approach, developed by economists and philosophers such as Amartya Sen and Martha Nussbaum [293, 235], provides a valuable theoretical framework for evaluating the empowering potential of AI systems. This approach focuses on expanding the real freedoms and opportunities that individuals have to live the lives they have reason to value. In the context of Human-AI Empowerment, the capability approach offers a normative framework for assessing the impact of AI interventions on human well-being and agency. It encourages the development of AI systems that not only enhance efficiency or productivity but also expand the range of meaningful choices available to individuals, thereby contributing to their overall empowerment and quality of life.

The concept of sociotechnical systems, originating from organizational theory and sociology, offers another important perspective on Human-AI Empowerment. This approach recognizes that technological systems are deeply intertwined with social structures, cultural norms, and organizational practices [328]. In the context of Human-AI Empowerment, the sociotechnical systems perspective highlights the need to consider the broader social and organizational contexts in which AI systems are deployed. It emphasizes that successful empowerment requires not only technological innovation but also appropriate social structures, governance mechanisms, and cultural adaptations to support the effective integration of AI into human activities.

The theoretical framework of embodied cognition, which posits that cognitive processes are deeply rooted in the body's interactions with the physical world, also has significant implications for Human-AI Empowerment. This perspective, developed by cognitive scientists and philosophers like Francisco Varela and Evan Thompson [338], challenges traditional notions of cognition as purely abstract information processing. In the context of Human-AI Empowerment, embodied cognition informs the design of AI interfaces that leverage natural human sensorimotor capabilities, such as gesture-based controls or augmented reality overlays. It also encourages the development of AI systems that can interpret and respond to embodied human cues, creating more intuitive and empowering human-AI interactions.

The concept of collective intelligence, which examines how groups of individuals can collaboratively solve problems and generate knowledge, provides another important theoretical lens for Human-AI Empowerment. This concept, explored by researchers like Thomas Malone [204], considers how the integration of human and AI can create synergies that surpass the capabilities of either alone. In the context of Human-AI Empowerment, collective intelligence frameworks inform the design of collaborative AI systems that facilitate group problem-solving, decision-making, and innovation. These systems leverage AI's data processing and pattern recognition capabilities while harnessing human creativity, contextual understanding, and social intelligence to achieve collective empowerment.

The theoretical framework of value alignment, prominent in AI ethics and safety research, is crucial for ensuring that Human-AI Empowerment efforts genuinely serve human interests and values. This framework, developed by AI researchers and ethicists like Stuart Russell [280], addresses the challenge of creating AI systems whose objectives and behaviors are aligned with human values and preferences. In the context of Human-AI Empowerment, value alignment theory informs the development of AI systems that not only enhance human capabilities but do so in ways that respect and promote human values, ethical principles, and long-term well-being.

Lastly, the concept of technological mediation, developed in the philosophy of technology by scholars like Don Ihde and Peter-Paul Verbeek [151, 339], offers valuable insights for Human-AI Empowerment. This perspective examines how technologies shape human perceptions, actions, and interpretations of the world. In the context of Human-AI Empowerment, technological mediation theory encourages critical reflection on how AI systems influence human experiences and decision-making processes. It informs the design of AI interfaces and interaction modalities that promote transparency, user agency, and critical engagement with AI-mediated information and recommendations.

These key concepts and theoretical frameworks collectively provide a rich intellectual foundation for the study and practice of Human-AI Empowerment. They offer diverse perspectives on the complex interplay between human cognition, social dynamics, and AI, informing the design of AI systems that genuinely enhance human capabilities and opportunities. As the field of Human-AI Empowerment continues to evolve, these foundational elements will undoubtedly be refined and expanded, incorporating new insights from ongoing research in AI, cognitive science, and related disciplines.

1.3 THE EVOLUTION OF HUMAN-CENTERED AI

The concept of Human-AI Empowerment represents the latest stage in the ongoing evolution of human-centered approaches to AI. This evolution reflects a growing recognition of the need to align AI development with human values, needs, and cognitive processes. The journey toward Human-AI Empowerment has been marked by significant shifts in perspectives on the role of AI in society and its relationship with human users. Understanding this historical context is crucial for appreciating the current state of Human-AI Empowerment and anticipating its future directions.

The roots of human-centered AI can be traced back to the early days of computing and cybernetics. Pioneers like Norbert Wiener, in his seminal work "The Human Use of Human Beings" (1950), emphasized the importance of designing technological systems that serve human needs and values [351]. This early recognition of the social implications of advanced computing systems laid the groundwork for subsequent human-centered approaches to technology development. As AI emerged as a distinct field in the 1950s and 1960s, researchers like John McCarthy and Marvin Minsky focused primarily on replicating human-like intelligence in machines. However, this early AI research often overlooked the complexities of human cognition and the potential for human-machine collaboration.

The 1980s saw a significant shift toward more human-centered approaches in AI and computing. The field of Human-Computer Interaction (HCI) emerged as a distinct discipline, focusing on designing computer systems that were usable, efficient, and satisfying for human users. Researchers like Ben Shneiderman advocated for "direct manipulation" interfaces that provided users with a sense of control and immediate feedback [297]. This period also saw the development of expert systems, which attempted to capture human expertise in specific domains. While these systems demonstrated the potential for AI to augment human capabilities, they also highlighted the challenges of effectively representing and utilizing human knowledge in machine-readable forms.

The 1990s and early 2000s witnessed a growing interest in adaptive and intelligent user interfaces. Researchers began exploring ways to create software that could learn from user behavior and adapt to individual preferences and needs. This period also saw the emergence of affective computing, pioneered by researchers like Rosalind Picard, which aimed to develop systems capable of recognizing and responding to human emotions [255]. These developments represented important steps toward more empathetic and human-centered AI systems, laying the groundwork for future Human-AI Empowerment approaches.

The concept of augmented intelligence gained prominence in the 2010s, shifting the focus from AI systems that compete with human intelligence to those that complement and enhance it. This perspective was championed by researchers and industry leaders who recognized the unique strengths of both human and AI. The IBM Watson system, which collaborated with human experts in fields like healthcare and financial services, exemplified this approach. Simultaneously, advances in machine learning and natural language processing enabled the development of more sophisticated virtual assistants and recommendation systems, further blurring the lines between human and AI.

Recent years have seen a growing emphasis on explainable AI (XAI) and interpretable machine learning. These approaches aim to create AI systems whose decision-making processes are transparent and understandable to human users. This trend reflects a recognition that true Human-AI

Empowerment requires not just powerful AI capabilities, but also the ability for humans to understand and trust the AI systems they interact with. Researchers like Cynthia Rudin have made significant contributions to developing interpretable machine learning models that maintain high performance while providing clear explanations for their predictions [279].

The emergence of Human-AI Empowerment as a distinct concept represents the culmination of these various threads in the evolution of human-centered AI. It builds on the foundational principles of human-centered design, the adaptive capabilities developed in intelligent user interfaces, the collaborative potential demonstrated by augmented intelligence approaches, and the transparency emphasized in explainable AI. However, Human-AI Empowerment goes beyond these predecessors by adopting a more holistic and long-term perspective on the relationship between humans and AI.

Human-AI Empowerment explicitly focuses on leveraging AI to enhance human capabilities and opportunities over extended periods, considering not just immediate task performance but also long-term personal and societal development. This long-term perspective distinguishes Human-AI Empowerment from earlier approaches that often focused on short-term usability or task efficiency. It recognizes that the true potential of AI lies not just in automating human tasks, but in creating symbiotic relationships that foster continuous human growth and adaptation.

The evolution toward Human-AI Empowerment has also been influenced by broader societal and ethical considerations surrounding AI development. As AI systems have become more powerful and pervasive, concerns about issues such as algorithmic bias, privacy, and the potential for technological unemployment have come to the forefront. These concerns have prompted a reevaluation of the goals and methods of AI development, emphasizing the need for approaches that prioritize human values and societal well-being. Human-AI Empowerment responds to these concerns by explicitly incorporating ethical considerations into the design and deployment of AI systems, aiming to create technologies that not only enhance individual capabilities but also contribute to broader social good. The shift toward Human-AI Empowerment has been further accelerated by advances in cognitive science and neuroscience, which have provided deeper insights into human cognition and learning processes. These insights have enabled the development of AI systems that more effectively complement and enhance human cognitive abilities. For example, research on the brain's plasticity and the mechanisms of skill acquisition has informed the design of AI-powered learning systems that adapt to individual cognitive styles and promote long-term skill development [76].

The COVID-19 pandemic has also played a role in accelerating the evolution toward Human-AI Empowerment. The rapid shift to remote work and online education highlighted the critical role of technology in enabling human activities and exposed the limitations of existing digital tools. This global experience has intensified interest in developing more sophisticated and empowering human-AI collaboration tools, capable of supporting complex cognitive tasks and facilitating remote collaboration across diverse domains.

As we look to the future, the evolution of Human-AI Empowerment is likely to continue, driven by ongoing advances in AI technologies and deepening understanding of human cognition and social dynamics. Emerging technologies such as brain-computer interfaces and augmented reality are opening new frontiers for human-AI interaction, potentially enabling even more intimate and empowering forms of collaboration. At the same time, growing awareness of the societal impacts of AI is likely to further emphasize the importance of human-centered approaches that prioritize empowerment, equity, and long-term human flourishing.

1.4 ETHICAL CONSIDERATIONS IN HUMAN-AI EMPOWERMENT

The pursuit of Human-AI Empowerment, while promising significant benefits for individuals and society, also raises a host of ethical considerations that must be carefully addressed. These ethical dimensions are not peripheral to the development of empowering AI systems, but are integral to their design, implementation, and governance. The complex interplay between artificial intelligence and human agency, capabilities, and values necessitates a thorough examination of the ethical implications at every stage of development and deployment. This section explores the key ethical considerations in Human-AI Empowerment, drawing on perspectives from philosophy, applied ethics, and science and technology studies. As illustrated in Figure 1.3, these ethical considerations permeate the entire AI empowerment lifecycle, from design to deployment and ongoing governance.

One of the primary ethical considerations in Human-AI Empowerment is the preservation and enhancement of human autonomy. While the goal of empowerment implies an increase in human capabilities and opportunities, there is a risk that excessive reliance on AI systems could paradoxically lead to a diminishment of human agency. This concern draws on philosophical debates about autonomy and authenticity, as articulated by thinkers like Charles Taylor and Harry Frankfurt [317][112]. In the context of Human-AI Empowerment, it raises questions about the extent to which AI systems should influence human decision-making and behavior. For instance, an AI system designed to enhance productivity might suggest optimal work schedules or task prioritizations, but taken to an extreme, such guidance could potentially undermine an individual's sense of self-direction and personal responsibility. Striking the right balance between AI assistance and human autonomy requires careful consideration of how AI recommendations are presented, the degree of human control over AI

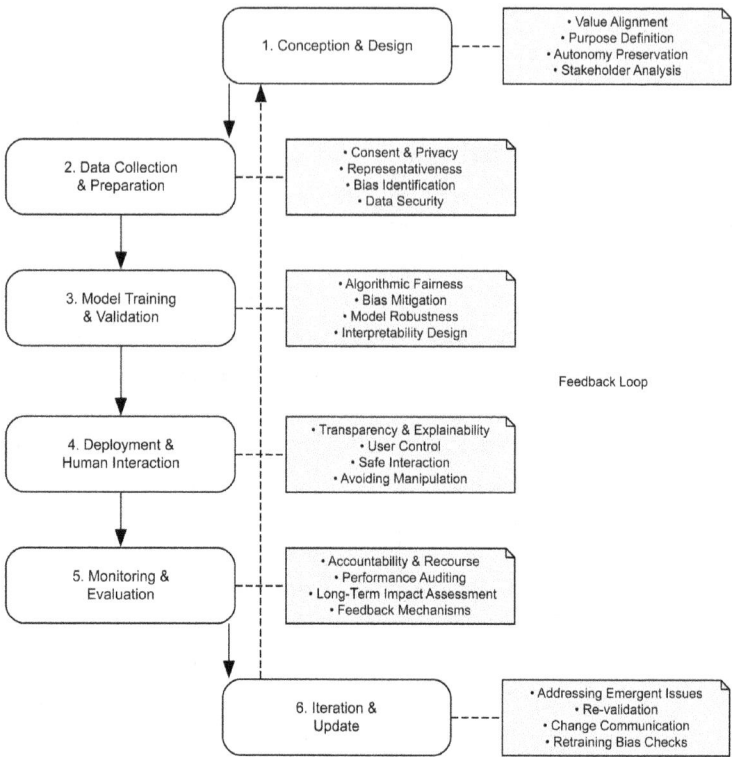

Figure 1.3 Ethical Considerations Throughout the AI Empowerment Lifecycle

systems, and the preservation of meaningful human choice in AI-mediated environments.

The issue of privacy and data ethics presents another crucial ethical dimension in Human-AI Empowerment. The development of AI systems capable of providing personalized assistance and adapting to individual needs often requires access to vast amounts of personal data. This data collection and usage raise significant privacy concerns, as articulated by privacy scholars like Helen Nissenbaum [231]. The ethical challenges extend beyond mere data protection to questions of informational self-determination and the right to cognitive privacy. As AI systems become more integrated into human cognitive processes, there is a need to establish ethical frameworks for managing the intimate knowledge these systems may accumulate about individual thoughts, preferences, and behaviors. These considerations necessitate the development of robust data governance mechanisms, transparent data usage policies, and technologies that enable fine-grained user control over personal data in Human-AI Empowerment systems.

The potential for bias and discrimination in AI systems represents a critical ethical concern in the context of Human-AI Empowerment. AI systems trained on historical data may perpetuate or even amplify existing societal biases, leading to unfair or discriminatory outcomes. This issue has been extensively documented in various AI applications, from facial recognition systems to hiring algorithms [51]. In the pursuit of Human-AI Empowerment, there is a risk that biased AI systems could exacerbate existing inequalities, providing greater benefits to already privileged groups while further marginalizing others. Addressing this ethical challenge requires a multifaceted approach, including diverse and representative data sets, algorithmic fairness techniques, and ongoing monitoring and auditing of AI systems for biased outcomes. Moreover, it necessitates a broader consideration of how AI systems can be designed to actively promote equity and inclusion, rather than merely avoiding discrimination.

The long-term societal impacts of widespread Human-AI Empowerment raise profound ethical questions about the future of human society and the nature of human flourishing. As AI systems become increasingly capable of enhancing human cognitive abilities and expanding individual opportunities, there is a need to consider the potential consequences for social structures, economic systems, and human relationships. Philosophers and ethicists have long debated the implications of human enhancement technologies, and many of these arguments apply to AI-enabled empowerment as well [45]. For instance, if Human-AI Empowerment leads to significant disparities in cognitive capabilities or economic opportunities, it could exacerbate social inequalities and potentially alter the very fabric of human society. These considerations necessitate ongoing ethical reflection and public dialogue about the values and goals that should guide the development and deployment of empowering AI technologies.

The issue of accountability and responsibility in Human-AI Empowerment systems presents another significant ethical challenge. As AI systems take on increasingly important roles in decision-making processes and task execution, questions arise about who bears responsibility for the outcomes of these human-AI collaborations. This issue becomes particularly complex in scenarios where the boundaries between human and AI contributions are blurred. Legal scholars and ethicists have grappled with similar questions in the context of autonomous systems, but Human-AI Empowerment introduces new dimensions to this debate [54]. For instance, if an AI system provides recommendations that significantly influence a human's decision-making process, how should responsibility be apportioned if that decision leads to harmful outcomes? Addressing these ethical challenges requires the development of new frameworks for understanding and allocating responsibility in human-AI collaborations, as well as appropriate legal and regulatory mechanisms to ensure accountability.

The potential for AI systems to influence human values and preferences raises ethical concerns about the preservation of human identity and cultural diversity. As AI systems become more deeply integrated into human cognitive processes and decision-making, there is a risk that they could subtly shape human values and goals in ways that may not be immediately apparent. This concern echoes broader debates in philosophy of technology about the non-neutrality of technological systems and their capacity to mediate human experiences and perceptions of the world [339]. In the context of Human-AI Empowerment, it raises questions about how to design AI systems that respect and preserve human values while still providing meaningful empowerment. This ethical challenge necessitates ongoing reflection on the values embedded in AI systems and the development of approaches that allow for human oversight and contestation of AI-mediated value propositions.

The ethical implications of cognitive offloading and AI dependence present another important consideration in Human-AI Empowerment. While AI systems can significantly enhance human cognitive capabilities, there is a risk that excessive reliance on these systems could lead to atrophy of certain human cognitive skills. This concern draws on research in cognitive psychology about the effects of technology use on human memory and problem-solving abilities [305]. In the pursuit of Human-AI Empowerment, it is crucial to consider how to design AI systems that enhance human cognition without undermining fundamental cognitive capabilities or creating problematic forms of dependence. This ethical challenge requires careful consideration of the appropriate balance between AI assistance and human cognitive effort, as well as the development of strategies to maintain and strengthen core human cognitive abilities in AI-rich environments.

The potential for AI systems to manipulate human behavior and decision-making, even when ostensibly aimed at empowerment, raises significant ethical concerns. The power of AI systems to process vast amounts of personal data and generate highly tailored recommendations creates the potential for subtle forms of influence that may not be immediately apparent to users. This issue connects to broader debates about digital manipulation and the ethics of persuasive technologies [312]. In the context of Human-AI Empowerment, it raises questions about the limits of acceptable influence and the importance of preserving human autonomy in AI-mediated environments. Addressing this ethical challenge requires the development of robust transparency mechanisms, user control features, and ethical guidelines for the design of AI systems that respect human agency and decision-making processes.

In conclusion, the ethical considerations in Human-AI Empowerment are multifaceted and deeply intertwined with fundamental questions about human nature, societal values, and the future of human-technology relations. Addressing these ethical challenges requires ongoing interdisciplinary dialogue, involving not only AI researchers and ethicists but also policymakers, social scientists, and members of the public. As the field of Human-AI Empowerment continues to evolve, it is crucial that ethical reflection remains at the forefront, guiding the development of AI systems that genuinely enhance human capabilities and opportunities while respecting human values, autonomy, and dignity.

II

Frameworks for Human-AI Empowerment

The study of Human-AI Empowerment necessitates a comprehensive and long-term approach to understanding the complex interactions between artificial intelligence systems and human development. Longitudinal studies offer a unique and valuable perspective on these interactions, allowing researchers to track changes in human capabilities, behaviors, and attitudes over extended periods of exposure to AI technologies [257]. This section explores the frameworks, methodologies, and challenges associated with conducting longitudinal studies in the field of Human-AI Empowerment. An examination of both quantitative and qualitative approaches, as well as mixed-method frameworks, aims to provide a robust foundation for researchers and practitioners seeking to assess the long-term impacts of AI on human empowerment [69].

2.1 DESIGNING LONGITUDINAL STUDIES FOR AI IMPACT

The design of longitudinal studies for assessing the impact of AI on human empowerment requires careful consideration of

DOI: 10.1201/9781003536628-2

numerous factors, including study duration, sample selection, data collection methods, and ethical considerations [217]. Longitudinal research, characterized by repeated observations of the same variables over long periods, offers unique insights into the developmental trajectories of human-AI interactions and their effects on individual and societal empowerment [302]. Figure 2.1 outlines the key stages and considerations involved in designing and executing these complex longitudinal investigations.

One of the primary considerations in designing longitudinal studies for AI impact is the determination of an appropriate time frame. The duration of the study must be sufficient to capture meaningful changes in human capabilities, behaviors, and attitudes resulting from sustained interaction with AI systems [30]. However, the rapid evolution of AI technologies presents a challenge, as the systems under study may become obsolete or significantly altered during the course of the research [7]. Researchers must therefore strike a balance between study duration and technological relevance. Some studies may opt for shorter durations of one to two years, focusing on specific AI applications and their immediate effects on human empowerment. Others may adopt longer time frames of five to ten years or more, aiming to capture broader trends and developmental patterns in human-AI interaction [295].

Sample selection and retention present another crucial aspect of longitudinal study design in the context of AI empowerment research [130]. The sample should be representative of the target population and sufficiently diverse to allow for meaningful analysis of different demographic groups and their experiences with AI systems. Researchers must consider factors such as age, gender, socioeconomic status, education level, and prior experience with technology when selecting participants [295]. Additionally, the sample size must be large enough to account for potential attrition over the course of the study, as participant dropout is a common challenge in longitudinal research. Strategies for minimizing attrition, such as regular communication with participants, provision of incentives, and

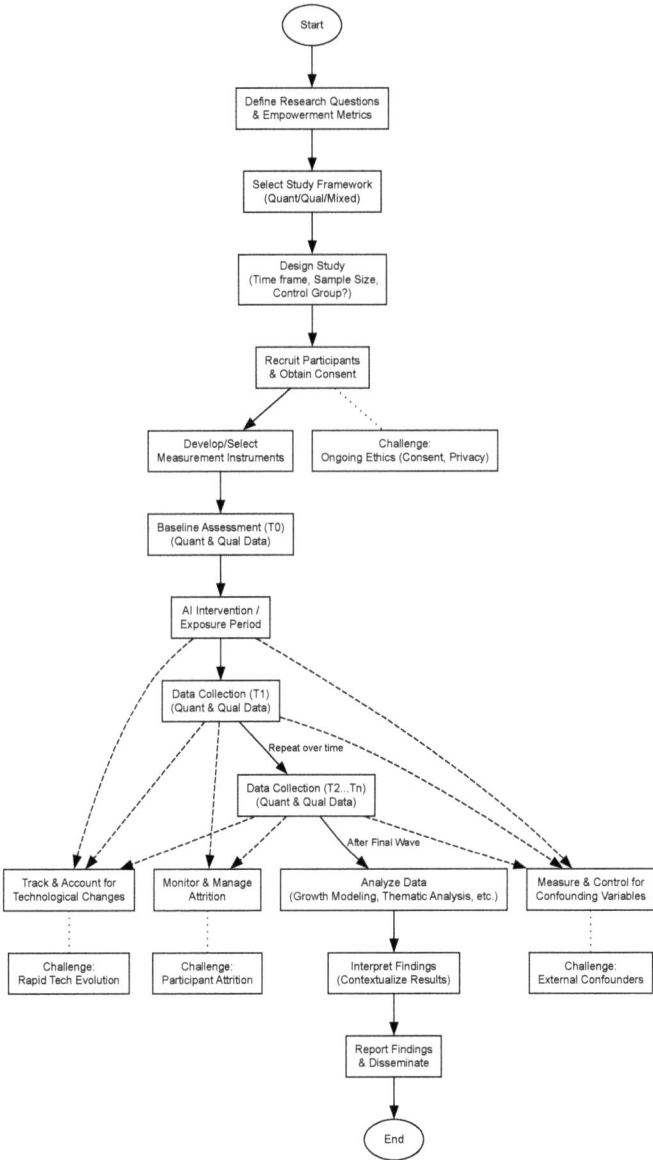

Figure 2.1 Designing and Executing Longitudinal Studies of AI Empowerment

flexible data collection methods, should be incorporated into the study design from the outset [271].

The selection of appropriate data collection methods is critical in longitudinal studies of AI empowerment [315]. A combination of quantitative and qualitative approaches is often necessary to capture the full complexity of human-AI interactions and their impacts on empowerment [69]. Quantitative methods may include standardized assessments of cognitive abilities, surveys measuring attitudes toward AI, and objective measures of task performance in AI-assisted environments. These methods allow for statistical analysis of trends and patterns over time. Qualitative methods, such as in-depth interviews, focus groups, and ethnographic observations, can provide rich, contextual data on participants' experiences, perceptions, and evolving relationships with AI systems [77].

The use of technology-mediated data collection methods presents both opportunities and challenges in longitudinal studies of AI empowerment [172]. Digital platforms and mobile applications can facilitate frequent and convenient data collection, allowing researchers to capture real-time information on participants' interactions with AI systems. Wearable devices and Internet of Things (IoT) sensors can provide continuous streams of behavioral and physiological data, offering insights into the subtle ways in which AI integration affects daily life [188]. However, the use of such technologies raises ethical concerns regarding privacy and data security, necessitating robust safeguards and transparent data management practices [222].

Ethical considerations must be at the forefront of longitudinal study design in AI empowerment research [156]. The extended nature of these studies, combined with the potentially intimate nature of human-AI interactions, requires careful attention to issues of informed consent, privacy protection, and participant well-being. Researchers must develop clear protocols for ongoing consent, allowing participants to reassess their willingness to continue in the study as it progresses [230]. Data anonymization and secure storage practices are essential to protect participant privacy, especially given the sensitive nature of long-term behavioral and cognitive data. Additionally, researchers must be prepared to address any negative impacts of AI interaction that may emerge during the study, potentially offering interventions or support to participants who experience adverse effects [348].

The design of longitudinal studies must also account for the potential impact of external factors on the observed effects of AI empowerment [88]. Societal changes, economic conditions, and technological advancements unrelated to the specific AI systems under study can all influence participants' experiences and outcomes. Researchers should incorporate methods for tracking and controlling for these external variables, such as including control groups or collecting data on relevant contextual factors [295]. This approach allows for a more accurate attribution of observed changes to the AI systems and interventions being studied.

Flexibility and adaptability are key principles in designing longitudinal studies of AI empowerment [257]. Given the rapid pace of technological change, study designs should incorporate mechanisms for adjusting research protocols and measurement tools as new AI capabilities emerge or as the focus of empowerment shifts. This may involve regular review and updating of survey instruments, the incorporation of new measurement technologies, or the addition of supplementary research questions to address emerging trends [7]. Building flexibility into the study design allows researchers to ensure that their investigations remain relevant and responsive to the evolving landscape of human-AI interaction.

2.2 QUANTITATIVE APPROACHES TO MEASURING EMPOWERMENT

Quantitative approaches play a crucial role in measuring and analyzing the impact of AI on human empowerment over time [69]. These methods provide structured, numerical data that allow for statistical analysis, trend identification, and the testing of hypotheses about the relationship between AI interaction and various aspects of human empowerment [98]. In the context of longitudinal studies, quantitative approaches offer the advantage of producing comparable data points across multiple time periods, enabling researchers to track changes and developments with precision.

One fundamental quantitative approach in longitudinal studies of AI empowerment is the use of standardized psychometric assessments [15]. These tools, which include cognitive tests, personality inventories, and measures of emotional intelligence, can provide objective measures of various aspects of human capabilities and traits. For example, cognitive assessments such as the Wechsler Adult Intelligence Scale (WAIS) or the Cambridge Brain Sciences battery can be used to track changes in specific cognitive domains like memory, attention, and problem-solving skills over the course of prolonged AI interaction [349][134]. Similarly, personality assessments like the Big Five Inventory can help researchers understand how sustained engagement with AI systems might influence personality traits such as openness to experience or conscientiousness [161]. Administering these assessments at regular intervals throughout a longitudinal study allows researchers to quantify changes in human cognitive and psychological characteristics that may be associated with AI empowerment.

Performance metrics derived from task-specific assessments offer another valuable quantitative approach to measuring empowerment [283]. These metrics can be tailored to the specific domains in which AI systems are expected to enhance human capabilities. For instance, in a study of AI-assisted learning, researchers might track improvements in test scores, problem-solving speed, or the complexity of tasks that participants can successfully complete [287]. In professional settings, performance metrics might include measures of productivity, error rates, or the quality of outputs in AI-augmented work processes. The key to effective use of performance metrics lies in designing assessments that are sensitive to the specific ways in which AI systems are hypothesized to empower human users, while also being generalizable enough to allow for comparisons across different AI applications and contexts. Survey instruments provide a versatile quantitative tool for assessing various dimensions of Human-AI Empowerment [111]. Likert-scale questionnaires can be used to measure attitudes toward AI, perceptions of self-efficacy in AI-mediated tasks, and subjective assessments of empowerment. For example, the AI Attitudes Scale developed by Schepman and Rodway offers a validated instrument for measuring public attitudes toward AI across multiple dimensions [286]. In longitudinal studies, repeated administration of such surveys can reveal shifts in perceptions and attitudes over time, providing insights into the psychological aspects of empowerment. Researchers can also develop custom survey instruments tailored to specific aspects of empowerment, such as perceived autonomy in AI-assisted decision-making or the sense of mastery over AI tools.

Digital trace data, collected through logs of user interactions with AI systems, present a rich source of quantitative information for longitudinal studies of empowerment [188]. This data can include metrics such as frequency and duration of AI system use, types of tasks performed with AI assistance, and patterns of user-initiated customization or override of AI recommendations. Analysis of digital trace data can reveal evolving patterns of human-AI interaction, providing objective measures of how individuals integrate AI tools into their daily activities and decision-making processes [178]. However, the use of such data requires careful consideration of privacy and ethical issues, as well as sophisticated data processing techniques to extract meaningful insights from large volumes of raw interaction logs.

Physiological measurements offer a unique quantitative perspective on the effects of AI interaction on human empowerment [53]. Techniques such as electroencephalography (EEG), functional magnetic resonance imaging (fMRI), and measurements of heart rate variability or skin conductance can provide objective data on cognitive load, stress levels, and emotional responses during AI-assisted tasks [83]. In longitudinal studies, these physiological markers can be tracked to understand how prolonged exposure to AI systems affects cognitive and emotional processes. For example, researchers might use EEG data to investigate whether long-term use of AI assistants leads to changes in neural patterns associated with problem-solving or decision-making. While powerful, these methods often require specialized equipment and expertise, and their application in naturalistic settings can be challenging.

Network analysis techniques offer quantitative tools for understanding the social dimensions of AI empowerment [43]. Mapping and analyzing patterns of communication, collaboration, and information flow within AI-augmented social networks allow researchers to quantify changes in social capital, knowledge dissemination, and collective problem-solving capabilities. Social network analysis metrics such as centrality measures, clustering coefficients, and network density can provide insights into how AI systems influence the structure and dynamics of human social interactions [188]. In longitudinal studies, these analyses can reveal how AI-mediated communication platforms and collaborative tools shape social empowerment over time.

Economic indicators provide another quantitative lens through which to examine AI empowerment [5]. Measures such as income levels, job mobility, and rates of entrepreneurship can be tracked over time to assess the economic dimensions of empowerment. For instance, longitudinal studies might investigate how exposure to AI-enhanced educational or professional development tools correlates with changes in earning potential or career advancement. Economic data can also be used to analyze broader societal trends related to AI empowerment, such as changes in income inequality or shifts in labor market dynamics [50].

The integration of multiple quantitative approaches through advanced statistical modeling techniques is crucial for developing a comprehensive understanding of AI empowerment [98]. Methods such as structural equation modeling, growth curve analysis, and multilevel modeling allow researchers to examine complex relationships between various quantitative measures over time. These techniques can help disentangle the effects of AI interaction from other factors influencing empowerment, and can reveal non-linear patterns or threshold effects in the empowerment process [302].

2.3 QUALITATIVE METHODS FOR ASSESSING HUMAN-AI INTERACTION

Qualitative methods play an indispensable role in assessing AI-Human interaction within the context of empowerment studies [77]. These approaches provide rich, contextual data that capture the nuanced experiences, perceptions, and evolving relationships between humans and AI systems. Qualitative methods are particularly valuable in longitudinal studies of Human-AI Empowerment, as they allow researchers to explore the subjective dimensions of empowerment that may be difficult to quantify [251]. This section examines key qualitative methodologies for assessing AI-Human interaction, discussing their applications, strengths, and considerations in the context of longitudinal research.

In-depth interviews stand as a cornerstone of qualitative research in Human-AI Empowerment studies [291]. These semi-structured or unstructured conversations allow researchers to delve deeply into participants' experiences, thoughts, and feelings about their interactions with AI systems. In longitudinal studies, repeated interviews with the same participants over time can reveal evolving perceptions, changing attitudes, and shifts in the nature of human-AI relationships [180]. For example, researchers might conduct annual interviews with individuals using AI-powered personal assistants, exploring how their sense of empowerment, dependency, or mastery changes over years of use. The flexibility of in-depth interviews allows for the exploration of unexpected themes or experiences that emerge during the course of the study, providing valuable insights into the complex and often unpredictable ways in which AI systems influence human empowerment.

Focus groups offer another valuable qualitative method for assessing AI-Human interaction in empowerment studies [179]. These facilitated group discussions can generate rich data on shared experiences, collective attitudes, and social norms surrounding AI use. In longitudinal research, recurring focus groups can track how community perceptions and social dynamics related to AI empowerment evolve over time [223]. For instance, a study might convene annual focus groups of teachers using AI-enhanced educational tools, exploring how their collective experiences and perspectives on student empowerment shift across multiple academic years. Focus groups are particularly useful for understanding the social and cultural dimensions of AI empowerment, as they allow researchers to observe how individuals construct meaning and negotiate their experiences with AI in a group context.

Ethnographic observation provides a powerful method for studying AI-Human interaction in naturalistic settings [133]. This approach involves immersive, long-term observation of individuals or communities as they interact with AI systems in their daily lives. In the context of empowerment studies, ethnographic methods can reveal subtle changes in behavior, decision-making processes, and social interactions that may not be captured through other means [256]. For example, a researcher might embed themselves in a workplace adopting AI-powered productivity tools, observing over months or years how employees' work practices, collaborations, and sense of professional empowerment evolve. Ethnographic approaches are particularly valuable for understanding the contextual factors that influence AI empowerment, as well as for identifying unintended consequences or emergent behaviors that arise from prolonged AI interaction.

Diary studies and self-reflection techniques offer a unique perspective on AI-Human interaction by allowing participants to document their experiences and thoughts in real-time [42]. In longitudinal research, participants might be asked to maintain regular journals or voice recordings about their interactions with AI systems, capturing immediate reactions, reflections, and perceived changes in their capabilities or opportunities. Digital platforms or mobile applications can facilitate this process, allowing for easy and frequent data collection [152]. These methods are particularly useful for capturing the day-to-day nuances of AI interaction that might be forgotten or overlooked in retrospective interviews. Additionally, the act of self-reflection itself can provide insights into how individuals construct meaning around their experiences with AI and conceptualize their own empowerment.

Visual methods, such as photo-elicitation or participatory video creation, offer innovative approaches to exploring the lived experience of AI interaction [135]. Participants might be asked to photograph or film aspects of their daily lives that relate to AI use and empowerment, providing visual narratives that complement verbal or written accounts. In longitudinal studies, these visual artifacts can serve as powerful tools for tracking changes in the physical and social environments surrounding AI use, as well as shifts in participants' perspectives and priorities over time [277]. Visual methods can be particularly effective in capturing aspects of AI empowerment that are difficult to articulate verbally, such as changes in the organization of physical spaces or non-verbal aspects of human-AI interaction.

Narrative analysis techniques can be applied to the qualitative data collected through interviews, diaries, or other methods to uncover deeper patterns and meanings in participants' accounts of AI empowerment [272]. Examining the stories people tell about their experiences with AI systems allows researchers to gain insights into how individuals construct their identities in relation to AI, how they perceive the role of AI in their personal development, and how they narratively frame the concept of empowerment itself. In longitudinal studies, tracking changes in narrative structures and themes over time can reveal evolving conceptualizations of human-AI relationships and shifting perceptions of agency and empowerment.

Critical incident technique (CIT) offers a focused approach to exploring significant events or experiences in AI-Human interaction that have particularly impactful effects on empowerment [103]. Participants are asked to recount specific incidents where AI interaction led to notable positive or negative outcomes in terms of their capabilities, opportunities, or sense of empowerment. In longitudinal research, the collection and analysis of critical incidents over time can provide insights into the key moments or tipping points in the empowerment process, as well as how individuals' criteria for judging significant AI interactions may evolve throughout their long-term engagement with AI systems [52].

Discourse analysis can be applied to qualitative data to examine how language use and communication patterns reflect and shape understandings of AI empowerment [119]. This method involves a close examination of the words, phrases, and rhetorical strategies used by participants when discussing their experiences with AI systems. In longitudinal studies, discourse analysis can reveal shifts in the linguistic framing of AI and empowerment over time, providing insights into changing societal narratives and individual conceptualizations of human-AI relationships [33]. For example, researchers might track how the metaphors used to describe AI systems evolve from tool-based analogies to more collaborative or even personified descriptions as individuals become more deeply engaged with AI technologies.

Grounded theory approaches offer a systematic method for developing theoretical frameworks of AI empowerment based on qualitative data [121]. This iterative process of data collection, coding, and analysis allows researchers to generate theories that are grounded in the lived experiences of participants. In longitudinal studies, grounded theory can be particularly valuable for understanding the process of empowerment as it unfolds over time, identifying key factors, stages, or conditions that influence the development of human capabilities through AI interaction [62]. The flexibility of grounded theory allows researchers to adapt their theoretical models as new data emerges, making it well-suited to studying the rapidly evolving field of AI empowerment.

Case studies provide an in-depth examination of specific instances of AI empowerment, offering a holistic view of the complex interplay between individual, technological, and contextual factors [363]. In longitudinal research, case studies might follow particular individuals, organizations, or communities over extended periods, documenting their journeys of AI adoption and empowerment. This approach allows for a rich exploration of the multifaceted nature of empowerment, considering factors such as organizational culture, individual personality traits, and broader societal influences [166]. Comparative case studies can be particularly illuminating, contrasting different trajectories of AI empowerment to identify key factors that contribute to successful outcomes.

Phenomenological approaches focus on understanding the lived experience of AI empowerment from the perspective of the individuals involved [120]. This method involves in-depth exploration of participants' subjective experiences, perceptions, and the meanings they attribute to their interactions with AI systems. In longitudinal studies, phenomenological research can reveal how the essence of the AI empowerment experience evolves over time, capturing shifts in individuals' sense of self, agency, and relationship to technology [195]. This approach is particularly valuable for understanding the existential and philosophical dimensions of long-term human-AI interaction.

The integration of multiple qualitative methods through triangulation can provide a more comprehensive understanding of AI-Human interaction in empowerment studies [132]. Combining data from interviews, observations, diaries, and other sources allows researchers to cross-validate findings and develop a more nuanced picture of the empowerment process. In longitudinal studies, methodological triangulation can help address the limitations of individual qualitative approaches and provide a more robust basis for tracking changes over time [201].

Ethical considerations are paramount in qualitative research on AI-Human interaction, particularly in longitudinal studies where researchers may develop long-term relationships with participants [74]. Issues of informed consent, confidentiality, and the potential for researcher influence on participants' experiences must be carefully addressed. Additionally, researchers must be sensitive to the power dynamics inherent in studying AI empowerment, particularly when working with vulnerable populations or in contexts where AI adoption may have significant implications for individuals' lives and livelihoods [156].

2.4 MIXED-METHOD FRAMEWORKS FOR COMPREHENSIVE EVALUATION

Mixed-method frameworks offer a powerful approach to comprehensively evaluating Human-AI Empowerment in longitudinal studies [69]. These frameworks integrate quantitative and qualitative methodologies, leveraging the strengths of each to provide a more nuanced and holistic understanding of the complex processes involved in AI-driven human empowerment [237]. Mixed-method frameworks combine the precision and generalizability of quantitative methods with the depth and contextual richness of qualitative approaches, enabling researchers to address the multifaceted nature of empowerment and capture its evolution over time. Figure 2.2 illustrates how quantitative and qualitative data streams can be integrated throughout the research process to achieve a more comprehensive evaluation.

The foundation of effective mixed-method frameworks lies in the principle of methodological complementarity [128]. This principle recognizes that quantitative and qualitative methods can address different aspects of the research question, providing a more comprehensive picture when used in combination. In the context of Human-AI Empowerment studies,

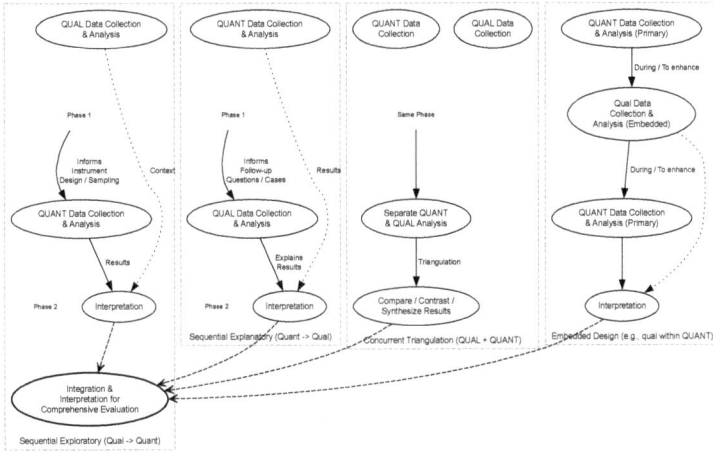

Figure 2.2 Integrating Quantitative and Qualitative Data in AI Empowerment Research

quantitative methods might be employed to measure changes in specific cognitive abilities or performance metrics, while qualitative methods could explore the subjective experiences and meaning-making processes associated with these changes [224]. For example, a longitudinal study of AI-assisted learning might combine standardized test scores and learning analytics (quantitative) with in-depth interviews and classroom observations (qualitative) to understand both the measurable outcomes and the lived experience of educational empowerment.

Sequential mixed-method designs offer one approach to integrating quantitative and qualitative methods in longitudinal studies of AI empowerment [69]. In this design, one methodological strand informs or builds upon the other in a sequential manner. For instance, an initial qualitative phase involving interviews and focus groups might be used to identify key dimensions of empowerment relevant to a specific AI application. These insights could then inform the development of quantitative instruments for measuring empowerment in subsequent phases of the study [70]. Conversely, quantitative analyses might reveal unexpected trends or correlations that are then explored in greater depth through qualitative inquiry. This iterative process allows for continual refinement of the research approach as the longitudinal study progresses.

Concurrent mixed-method designs, where quantitative and qualitative data are collected and analyzed simultaneously, can provide real-time triangulation of findings in longitudinal AI empowerment studies [343]. This approach allows researchers to corroborate findings across methods, identify discrepancies that merit further investigation, and develop a more comprehensive understanding of empowerment processes as they unfold. For example, participants might complete regular quantitative assessments of their AI interaction patterns and perceived empowerment, while simultaneously maintaining qualitative diaries or participating in periodic in-depth interviews [8]. The concurrent analysis of these data streams can provide insights into how objective measures of AI use and performance align with subjective experiences of empowerment.

Embedded mixed-method designs involve nesting one methodological approach within a larger study dominated by the other method [69]. In longitudinal studies of AI empowerment, this might take the form of a primarily quantitative study that incorporates periodic qualitative components to provide context and explanation for observed trends. For instance, a large-scale survey tracking changes in AI attitudes and self-efficacy over time might include a subset of participants who engage in more intensive qualitative research, such as ethnographic observation or repeated in-depth interviews. This approach allows for broad, generalizable findings while still providing rich, contextual data on the mechanisms of empowerment.

The integration of data from mixed-method studies poses both challenges and opportunities in longitudinal AI empowerment research [35]. Techniques such as data transformation, where qualitative data are quantified or quantitative data are qualitized, can facilitate the comparison and synthesis of findings across methods. For example, thematic analysis of interview data might be used to develop quantitative codes for analyzing larger text corpora, or statistical clusters derived from quantitative data might be used to structure qualitative inquiry into different empowerment trajectories [246]. Advanced analytical techniques such as qualitative comparative analysis (QCA) or mixed-methods matrix analysis can help researchers identify patterns and relationships across diverse data types [266].

Longitudinal case studies using mixed methods offer a particularly powerful approach to understanding the complex dynamics of AI empowerment [202]. Following specific individuals, organizations, or communities over extended periods and employing a range of quantitative and qualitative techniques enable researchers to develop rich, contextually grounded accounts of the empowerment process. These case studies might combine quantitative measures of cognitive performance and AI system use with qualitative explorations of changing perceptions, social dynamics, and environmental factors. The longitudinal nature of these studies allows for the identification of critical junctures, feedback loops, and non-linear developments in the empowerment process that might be missed in cross-sectional research [66].

The development of integrated theoretical frameworks is a key outcome of mixed-method longitudinal studies in AI empowerment [69]. Synthesizing insights from quantitative and qualitative strands of research allows researchers to construct more comprehensive and nuanced models of how AI interaction leads to human empowerment over time. These frameworks might incorporate both measurable indicators of empowerment and conceptual models of the underlying psychological and social processes. The iterative refinement of these frameworks over the course of longitudinal studies can contribute significantly to the theoretical understanding of Human-AI Empowerment [316].

Ethical considerations in mixed-method frameworks for AI empowerment studies require careful attention, particularly in longitudinal research [156]. The combination of quantitative and qualitative approaches may involve collecting a wide range of personal data over extended periods, raising concerns about privacy, confidentiality, and the potential for unintended uses of research findings. Researchers must develop robust protocols for data protection and participant consent that account for the evolving nature of AI technologies and the potential for unforeseen ethical challenges to emerge over the course of the study [46].

The selection and training of research teams for mixed-method longitudinal studies of AI empowerment present unique challenges [69]. Researchers must be skilled in both quantitative and qualitative methodologies, as well as possessing domain expertise in AI technologies and human development. Interdisciplinary collaboration is often essential, bringing together experts from fields such as computer science, psychology, sociology, and ethics. The long-term nature of these studies also requires strategies for maintaining team cohesion and methodological consistency over extended periods, even as research personnel may change [257].

2.5 CHALLENGES AND LIMITATIONS IN LONGITUDINAL AI STUDIES

Longitudinal studies of Human-AI Empowerment, while offering valuable insights into the long-term effects of AI on human capabilities and opportunities, face a number of significant challenges and limitations [257]. These issues stem from the complex, dynamic nature of AI technologies, the multifaceted concept of empowerment, and the practical difficulties inherent in conducting extended research projects. Understanding these challenges is crucial for researchers designing and implementing longitudinal studies, as well as for those interpreting and applying their findings.

One of the primary challenges in longitudinal AI studies is the rapid pace of technological change [7]. AI systems are continually evolving, with new capabilities and applications emerging at a rapid rate. This creates a moving target for researchers, as the AI technologies being studied at the beginning of a longitudinal project may be significantly different or even obsolete by its conclusion. This challenge is compounded by the fact that AI systems often undergo frequent updates and modifications, potentially altering their impact on users over the course of the study [50]. Researchers must grapple with questions of how to maintain consistency in their measures of AI interaction and empowerment while also accounting for technological advancements. Strategies such as regularly updating assessment tools, incorporating flexibility into research protocols, and carefully documenting changes in AI systems can help address this challenge, but the fundamental tension between study longevity and technological currency remains.

The potential for selection bias and attrition presents another significant challenge in longitudinal AI studies [130]. Participants who choose to engage in long-term research on AI empowerment may not be representative of the broader population, potentially skewing results toward those who are more technologically inclined or have more positive attitudes toward AI. Moreover, the extended nature of longitudinal studies increases the risk of participant dropout, which can introduce bias if attrition is non-random [131]. For example, individuals who feel less empowered by AI systems may be more likely to discontinue participation, leading to an overestimation of positive empowerment effects. Researchers must employ strategies such as diverse recruitment methods, incentives for continued participation, and statistical techniques for handling missing data to mitigate these issues. However, the challenge of maintaining a representative sample over extended periods remains a significant limitation of longitudinal AI studies.

The complexity of isolating the effects of AI interaction from other factors influencing human empowerment poses a

major challenge in longitudinal research [295]. Over the course
of a long-term study, participants will inevitably experience a
wide range of life events, educational experiences, and societal
changes that could impact their sense of empowerment inde-
pendently of their AI interactions. Disentangling these various
influences requires sophisticated research designs and analytical techniques [88]. Control groups, matched comparisons, and
statistical methods such as propensity score matching can help
address this challenge, but the difficulty of establishing clear
causal relationships in complex, real-world settings remains a
limitation of longitudinal AI studies.

Ethical considerations present ongoing challenges in longitudinal AI research, particularly as the nature of AI technologies and their societal implications may change over the course
of the study [222]. Issues of informed consent become complex when the long-term effects of AI interaction are not fully
known at the study's outset. Researchers must grapple with
questions of how to balance the need for consistent methodology with the ethical imperative to protect participants from
unforeseen risks. Additionally, the collection of extensive personal data over long periods raises significant privacy concerns, requiring robust data protection measures and ongoing
ethical review [230]. The potential for research findings to influence AI development and policy also raises ethical questions
about the responsibilities of researchers to their participants
and to society at large.

The multidimensional nature of empowerment poses conceptual and methodological challenges in longitudinal AI
studies [368]. Empowerment encompasses various aspects of
human experience, including cognitive abilities, emotional
well-being, social relationships, and economic opportunities.
Developing comprehensive measures that capture all these dimensions while remaining sensitive to change over time is a
significant challenge [236]. Moreover, the meaning and manifestation of empowerment may itself evolve over the course
of a longitudinal study, both at the individual and societal level. Researchers must grapple with how to maintain

conceptual consistency while also accounting for shifting understandings of empowerment in response to technological and social changes.

The potential for researcher bias and the influence of funding sources present challenges to the objectivity and credibility of longitudinal AI studies [155]. Researchers' own attitudes toward AI and preconceptions about its empowering potential may influence study design, data interpretation, and the framing of results. Additionally, the significant resources required for long-term research often necessitate funding from industry or other stakeholders with vested interests in AI development. This can raise questions about the independence of the research and the potential for findings to be influenced by funders' agendas [1]. Transparency in research methods, preregistration of study protocols, and diverse funding sources can help address these concerns, but the challenge of maintaining both actual and perceived objectivity in a field with significant commercial and political implications remains.

The generalizability of findings from longitudinal AI studies is limited by the specific contexts in which they are conducted [295]. AI systems and their effects on empowerment may vary significantly across cultural, socioeconomic, and technological contexts. A study conducted in a technologically advanced, urban environment may yield very different results from one carried out in a rural or developing world context. Moreover, the rapid pace of AI development means that findings from even recently completed longitudinal studies may not be fully applicable to current or future AI systems. Researchers must be cautious in extrapolating results beyond the specific populations and technologies studied, and should strive to conduct comparative studies across diverse contexts.

The practical challenges of maintaining consistent research methodologies and team expertise over extended periods pose significant hurdles in longitudinal AI studies [257]. Changes in research personnel, evolving ethical standards, and shifts in funding priorities can all impact the consistency and continuity of long-term research projects. Ensuring that data collection and analysis methods remain consistent while also incorporating methodological advancements is a delicate balance. Additionally, maintaining a research team with the necessary interdisciplinary expertise in AI, psychology, ethics, and other relevant fields over many years can be challenging. In conclusion, while longitudinal studies offer invaluable insights into the long-term effects of AI on human empowerment, they face numerous challenges and limitations. The rapid pace of technological change, issues of sample bias and attrition, the complexity of isolating AI effects, ethical considerations, conceptual challenges in measuring empowerment, potential biases, limited generalizability, and practical difficulties in maintaining long-term research projects all pose significant hurdles [257]. Acknowledging and addressing these challenges is crucial for conducting rigorous, meaningful research on Human-AI Empowerment. Researchers can contribute to a more nuanced and comprehensive understanding of how AI technologies shape human capabilities and opportunities over time through the development of innovative research designs, the employment of diverse methodologies, and the maintenance of a critical awareness of the limitations of their work [295].

To address these challenges, researchers in the field of Human-AI Empowerment must adopt innovative approaches and constantly refine their methodologies. One promising strategy is the development of adaptive research designs that can evolve alongside the AI technologies being studied [258]. These designs might incorporate flexible measurement protocols that can be updated to reflect technological advancements while maintaining comparability with earlier data points. Additionally, the use of AI-powered research tools, such as natural language processing for analyzing qualitative data or machine learning algorithms for identifying patterns in complex longitudinal datasets, may help researchers keep pace with the rapid evolution of AI technologies [203].

Collaboration between academic researchers and industry partners can provide valuable opportunities for conducting longitudinal studies in real-world settings, but must be carefully managed to maintain scientific integrity and ethical standards [38]. Developing clear guidelines for industry-academic partnerships, including provisions for data sharing, publication rights, and participant protection, can help ensure that such collaborations contribute to the advancement of knowledge while safeguarding the interests of study participants and the broader public.

The challenges of longitudinal AI studies also underscore the importance of developing robust theoretical frameworks that can guide research across diverse contexts and technological landscapes [361]. These frameworks should be flexible enough to accommodate the dynamic nature of AI technologies while providing a stable foundation for comparing and synthesizing findings across different studies and time periods. Interdisciplinary efforts to integrate insights from computer science, cognitive psychology, sociology, and other relevant fields can contribute to the development of more comprehensive and resilient theoretical models of Human-AI Empowerment.

As the field of Human-AI Empowerment research matures, the establishment of large-scale, collaborative research initiatives may help address some of the challenges associated with longitudinal studies [366]. These initiatives could pool resources and expertise across multiple institutions, enabling longer-term studies with larger and more diverse participant pools. Such collaborations could also facilitate the creation of shared data repositories and standardized measurement protocols, enhancing the comparability and generalizability of findings across different research projects.

The ethical challenges posed by longitudinal AI studies call for ongoing dialogue and collaboration between researchers, ethicists, policymakers, and members of the public [222]. Developing ethical frameworks that can adapt to the evolving capabilities and implications of AI technologies, while maintaining core principles of participant protection and societal benefit, is crucial. This may involve the creation of ethical review processes specifically tailored to longitudinal AI research, as well as mechanisms for ongoing stakeholder engagement throughout the course of long-term studies.

Despite these challenges, the potential insights offered by longitudinal studies of Human-AI Empowerment make them an essential component of our efforts to understand and shape the future of human-AI interaction. Tracking the long-term effects of AI on human capabilities, attitudes, and social structures allows these studies to inform the development of AI technologies that genuinely enhance human potential and contribute to individual and societal well-being [298]. Moreover, the findings from such research can play a crucial role in guiding policy decisions and public discourse around the responsible development and deployment of AI systems.

In conclusion, while longitudinal studies of Human-AI Empowerment face significant challenges, they also offer unparalleled opportunities to deepen our understanding of the complex, evolving relationship between humans and AI technologies. Acknowledging and actively addressing the limitations and ethical considerations inherent in this type of research enables scholars to contribute to a more nuanced, comprehensive, and ethically grounded body of knowledge about the long-term impacts of AI on human empowerment. This knowledge, in turn, can inform the development of AI systems that truly serve to enhance human capabilities, expand opportunities, and contribute to the flourishing of individuals and societies in an increasingly AI-mediated world [280].

III

Strategies for Empowering Humans Through AI Collaboration

The development of effective strategies for empowering humans through AI collaboration is a critical challenge in the rapidly evolving landscape of artificial intelligence. This section explores various approaches and methodologies aimed at fostering a mutually beneficial synergy between humans and AI systems. An examination of AI Empowerment approaches, Human-Computer Interaction (HCI) methods for long-term engagement, and insights from various disciplines on long-term goal management aims to provide a comprehensive framework for designing and implementing AI systems that genuinely enhance human capabilities and opportunities over extended periods.

3.1 AI EMPOWERMENT APPROACHES

AI Empowerment approaches focus on developing AI systems that actively contribute to enhancing human capabilities and

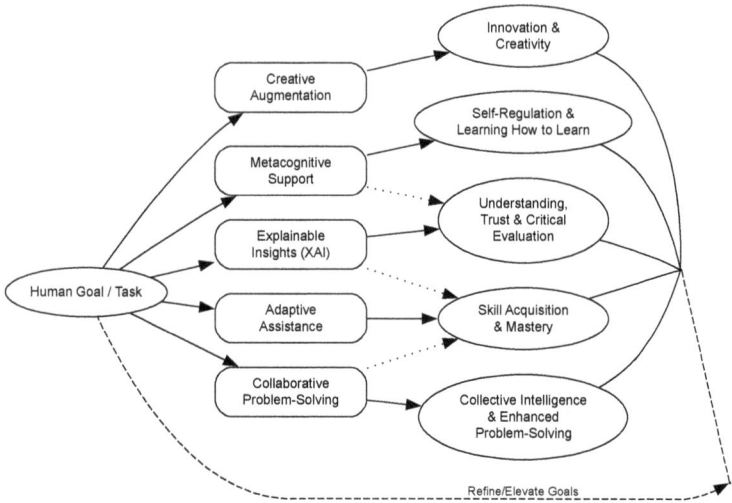

Figure 3.1 AI-Driven Strategies for Human Capability Enhancement

expanding opportunities for personal and professional growth. These approaches are grounded in the principle that AI should serve as a tool for augmenting human intelligence rather than replacing it, aligning with the concept of intelligence amplification proposed by Douglas Engelbart [91]. The goal is to create AI systems that not only assist in task completion but also foster the development of new skills, knowledge, and problem-solving abilities in their human collaborators. Figure 3.1 depicts several key AI-driven strategies aimed at enhancing human capabilities.

One key strategy in AI Empowerment is the development of adaptive AI systems that can tailor their support to individual users' needs, preferences, and learning styles [357]. These systems employ machine learning algorithms to analyze user behavior, performance, and feedback, continuously adjusting their interactions to provide optimal support. For example, an AI-powered educational platform might adapt its content presentation, pacing, and difficulty level based on a student's progress and engagement patterns, thereby facilitating more effective learning outcomes [12]. This personalized approach to AI assistance can help users build confidence, develop new competencies, and tackle increasingly complex challenges over time.

Another important aspect of AI Empowerment is the design of transparent and explainable AI systems [129]. These systems provide clear insights into their decision-making processes and reasoning, enabling users to understand, critically evaluate, and learn from AI-generated recommendations or solutions. This transparency not only builds trust but also supports the development of human expertise by allowing users to gain insights from the AI's analytical approach. For instance, an AI system used in medical diagnosis might not only provide a recommendation but also explain the key factors and reasoning behind its conclusion, thereby enhancing the physician's diagnostic skills and knowledge [144].

Collaborative AI systems that actively involve humans in the problem-solving process represent another powerful approach to empowerment [165]. These systems are designed to leverage the complementary strengths of human and artificial intelligence, creating a symbiotic relationship that enhances overall performance and fosters mutual learning. For example, in the field of scientific research, AI systems can assist human scientists by generating hypotheses, analyzing vast datasets, and identifying patterns, while humans provide creative insights, contextual understanding, and critical evaluation of the AI's outputs [175]. This collaborative approach not only leads to more effective problem-solving but also helps humans develop new analytical skills and domain knowledge through their interaction with the AI system.

The concept of AI as a creativity enhancer is another important strategy in empowerment approaches [199]. AI systems can be designed to stimulate human creativity by generating novel ideas, making unexpected connections, or providing alternative perspectives on a problem. For instance, AI-powered tools in fields such as design, music composition, or writing can suggest unconventional combinations or variations, challenging human creators to explore new creative directions and expand their artistic capabilities [113]. Serving as a source of inspiration and a tool for rapid prototyping, these AI systems can help humans push the boundaries of their creative potential.

Empowerment through AI also involves the development of systems that support metacognitive skills and self-regulated learning [27]. These AI tools can help users become more aware of their own thinking processes, learning strategies, and problem-solving approaches. For example, an AI system might provide feedback not just on task outcomes but also on the user's approach to the task, suggesting more effective strategies or pointing out potential biases in their thinking. This metacognitive support can foster the development of critical thinking skills, self-reflection, and lifelong learning capabilities [136].

The integration of AI systems into collaborative work environments presents another avenue for human empowerment [290]. AI-enhanced collaboration tools can facilitate more effective team communication, task coordination, and knowledge sharing. These systems might analyze team dynamics, suggest optimal task allocations based on individual strengths, or identify potential synergies between team members' ideas. These AI tools enhance collective intelligence and facilitate more efficient collaboration, empowering individuals to contribute more effectively to team goals and develop their collaborative skills [205].

Ethical considerations play a crucial role in AI Empowerment approaches [106]. It is essential to design AI systems that not only enhance human capabilities but also align with human values, respect individual autonomy, and promote social good. This involves careful consideration of issues such as privacy, fairness, and the potential for unintended consequences. For instance, AI systems designed for empowerment should avoid creating dependencies or undermining human agency. Instead, they should be developed with the goal of gradually increasing user autonomy and capability over time [267].

In conclusion, AI Empowerment approaches encompass a wide range of strategies aimed at creating AI systems that actively contribute to human growth and development. These approaches focus on adaptivity, transparency, collaboration, creativity enhancement, metacognitive support, and ethical considerations, seeking to harness the power of AI to expand human potential and create a more empowering human-AI relationship. As AI technologies continue to advance, the development and refinement of these empowerment strategies will play a crucial role in shaping a future where AI serves as a powerful tool for human flourishing.

3.2 HUMAN-COMPUTER INTERACTION METHODS FOR LONG-TERM ENGAGEMENT

HCI methods for long-term engagement are crucial in designing AI systems that not only empower users in the short term but also foster sustained growth and development over extended periods. These methods focus on creating interfaces and interaction paradigms that maintain user interest, facilitate ongoing learning, and adapt to changing user needs and capabilities over time. Long-term engagement is particularly important in the context of AI empowerment, as the full potential of human-AI collaboration often emerges through prolonged interaction and co-evolution of both human and AI capabilities [158].

One key approach in designing for long-term engagement is the implementation of progressive challenge and skill development [71]. This concept, rooted in flow theory, suggests that optimal engagement occurs when the level of challenge matches the user's skill level, gradually increasing in complexity as the user's capabilities grow. In the context of AI-empowered systems, this might involve dynamically adjusting the complexity of tasks or the level of AI assistance based on the user's progress. For example, an AI-powered language learning application might gradually introduce more complex grammatical structures and vocabulary, while reducing explicit guidance, as the user's proficiency improves [122]. This approach not only maintains user engagement but also promotes continuous skill development and a sense of accomplishment.

Gamification elements can be effectively integrated into AI systems to enhance long-term engagement and motivation [78]. Incorporating game-like features such as points, badges, leaderboards, and narrative elements allows AI-empowered applications to create a more engaging and rewarding user experience. However, it's crucial to design these elements thoughtfully to ensure they promote meaningful engagement rather than shallow interaction. For instance, an AI system for professional development might use gamification to reward not just task completion but also the application of new skills, creative problem-solving, or knowledge sharing with peers [186]. When combined with AI's ability to personalize experiences, gamification can be tailored to individual motivational profiles, maximizing its effectiveness in promoting long-term engagement.

Adaptive user interfaces that evolve based on user behavior and preferences play a significant role in maintaining long-term engagement [117]. These interfaces use machine learning algorithms to analyze user interactions over time, adjusting their layout, functionality, and information presentation to better suit individual needs and work patterns. For example, an AI-powered productivity suite might reorganize its interface to prioritize frequently used features, suggest personalized shortcuts, or adjust its visual design based on the user's cognitive load and work context. Continually optimizing the user experience allows adaptive interfaces to reduce friction, increase efficiency, and maintain user interest over extended periods [99].

The incorporation of social elements and community features can significantly enhance long-term engagement with AI systems [260]. Facilitating connections between users, sharing of experiences, and collaborative learning enables these social components to create a sense of belonging and mutual support. For instance, an AI-empowered health and wellness platform might include community forums, peer challenges, or mentor-mentee matching systems to encourage users to share their progress, seek advice, and motivate each other. The AI system can play a role in facilitating these social interactions by suggesting relevant connections, moderating discussions, or providing personalized community content recommendations [273].

Narrative and contextual framing of AI interactions can contribute to sustained engagement by providing meaning and relevance to the user's experiences [138]. Rather than presenting AI assistance as a series of isolated tasks or recommendations, framing these interactions within a larger narrative or personal growth journey can increase user investment and motivation. For example, an AI career development system might frame its interactions in terms of a hero's journey, with the user progressing through various challenges and milestones on their path to professional mastery. This narrative approach can help users see the long-term value of their engagement with the AI system and maintain motivation through setbacks or plateaus [215].

Incorporating elements of surprise and serendipity into AI system interactions can help maintain user interest and prevent engagement fatigue over time [16]. While consistency and predictability are important for usability, occasional unexpected or delightful interactions can reignite user curiosity and enthusiasm. This might involve the AI system introducing novel challenges, presenting information from unexplored domains, or making creative connections between seemingly unrelated concepts. For instance, a personal AI assistant might occasionally suggest a book, article, or learning opportunity slightly outside the user's usual interests, potentially sparking new areas of growth or insight [168].

Transparency and user control are essential elements in fostering long-term trust and engagement with AI systems [298]. Users should have clear visibility into how the AI is making decisions, what data it's using, and how their interactions are shaping the system's behavior. Moreover, providing users with the ability to customize, override, or refine the AI's operations can promote a sense of agency and partnership. For example, an AI system for financial planning might allow users to adjust risk tolerance parameters, exclude certain types of investments, or provide feedback on recommendations, ensuring that the AI's assistance aligns with the user's values and goals over time [2].

Regular reflection and progress tracking features can enhance long-term engagement by helping users recognize their growth and the value of their interaction with the AI system [192]. These features might include visualizations of skill development over time, summaries of key learnings or achievements, or prompted self-reflection exercises. Making progress tangible and encouraging metacognition allows these elements to reinforce the user's commitment to ongoing engagement and development. For instance, an AI-powered creative writing assistant might provide periodic analyses of the user's evolving writing style, complexity, and thematic exploration, helping the user appreciate their artistic growth [167].

Finally, the design of graceful degradation and re-engagement mechanisms is crucial for maintaining long-term relationships with AI systems [170]. Users may go through periods of reduced engagement due to various life circumstances, and the system should be designed to handle these fluctuations smoothly. This might involve adjusting the frequency and nature of interactions during low-engagement periods, providing easy on-ramps for users returning after a break, and ensuring that the system maintains relevant assistance even with sporadic use. Accommodating the natural ebbs and flows of user engagement enables AI systems to foster more sustainable long-term relationships and continue to provide value over extended periods [32].

In conclusion, HCI methods for long-term engagement in AI-empowered systems involve a multifaceted approach that combines progressive challenge, gamification, adaptive interfaces, social elements, narrative framing, elements of surprise,

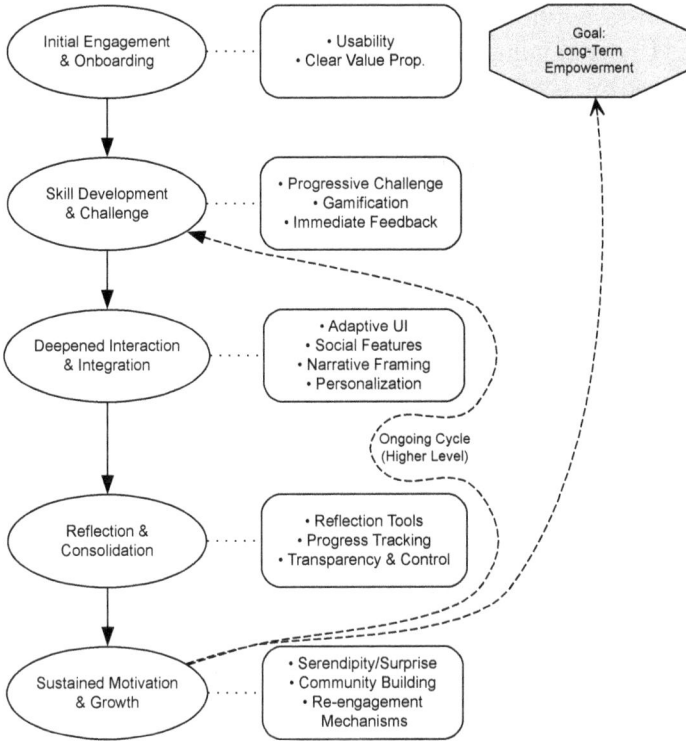

Figure 3.2 Sustaining Engagement for Long-Term AI Empowerment

transparency, reflection tools, and flexible engagement models. Thoughtfully integrating these elements allows designers to create AI systems that not only capture initial user interest but also foster sustained engagement, continued growth, and lasting empowerment over extended periods of human-AI collaboration. Figure 3.2 provides a visual summary of these integrated strategies for sustaining long-term engagement.

3.3 PSYCHOLOGICAL INSIGHTS ON LONG-TERM GOAL MANAGEMENT

Psychological insights into long-term goal management provide crucial foundations for designing AI systems that effectively support human empowerment over extended periods. Understanding the cognitive and motivational processes involved in setting, pursuing, and achieving long-term goals can inform the development of AI tools that align with and enhance these natural human tendencies. This section explores key psychological theories and findings relevant to long-term goal management and their implications for AI-empowered systems.

Goal-Setting Theory, developed by Locke and Latham, offers fundamental insights into effective goal management [197]. The theory posits that specific, challenging goals lead to higher performance than vague or easy goals. In the context of AI empowerment, this suggests that systems should encourage users to set clear, ambitious, yet achievable objectives. AI can play a role in helping users break down long-term goals into specific, measurable sub-goals and providing real-time feedback on progress. For instance, an AI career development platform might assist users in defining precise career milestones and generate personalized action plans with concrete, challenging steps toward these goals [187].

The concept of Implementation Intentions, introduced by Gollwitzer, highlights the importance of planning specific actions in advance to achieve goals [124]. This approach involves creating "if-then" plans that link anticipated situational cues with goal-directed responses. AI systems can leverage this insight by helping users create and remember these implementation intentions. For example, a health and wellness AI might assist users in formulating specific plans like "If it's 7 AM, then I'll do a 15-minute yoga session," and provide timely reminders or adaptive suggestions based on the user's context and behavior patterns [6].

Self-Determination Theory (SDT), developed by Deci and Ryan, emphasizes the importance of autonomy, competence, and relatedness in fostering intrinsic motivation and well-being [281]. AI systems designed for long-term empowerment should strive to support these basic psychological needs. This might involve providing users with meaningful choices in their goal pursuit (autonomy), offering optimal challenges and positive feedback (competence), and facilitating connections with others pursuing similar goals (relatedness). An AI learning platform, for instance, could allow users to choose their learning paths, provide adaptive challenges, and create study groups with peers, thereby supporting all three aspects of SDT [63]. Figure 3.3 illustrates how AI support can be integrated into the various stages of the goal management process, drawing on these psychological insights.

The Transtheoretical Model of Behavior Change, proposed by Prochaska and DiClemente, describes the stages individuals go through when changing behaviors: precontemplation, contemplation, preparation, action, and maintenance [263]. AI systems can use this model to tailor their support based on the user's current stage of change. For example, a financial planning AI might provide educational content for users in the contemplation stage, assist with concrete planning for those in the preparation stage, and offer reinforcement and relapse prevention strategies for users in the maintenance stage of a new saving behavior [233].

Cognitive Load Theory, developed by Sweller, emphasizes the limited capacity of working memory and its implications for learning and problem-solving [313]. AI systems can apply this theory by managing the cognitive load associated with long-term goal pursuit. This might involve breaking complex tasks into manageable chunks, providing just-in-time information to reduce extraneous cognitive load, and gradually increasing intrinsic cognitive load as users develop expertise. An AI writing assistant, for example, could scaffold the writing process by suggesting outlines, providing relevant research at appropriate moments, and gradually reducing explicit guidance as the user's skills improve [248].

The concept of Growth Mindset, introduced by Dweck, highlights the importance of believing that abilities can be developed through effort and learning [86]. AI systems can promote a growth mindset by framing challenges as opportunities for learning, providing process-focused feedback, and

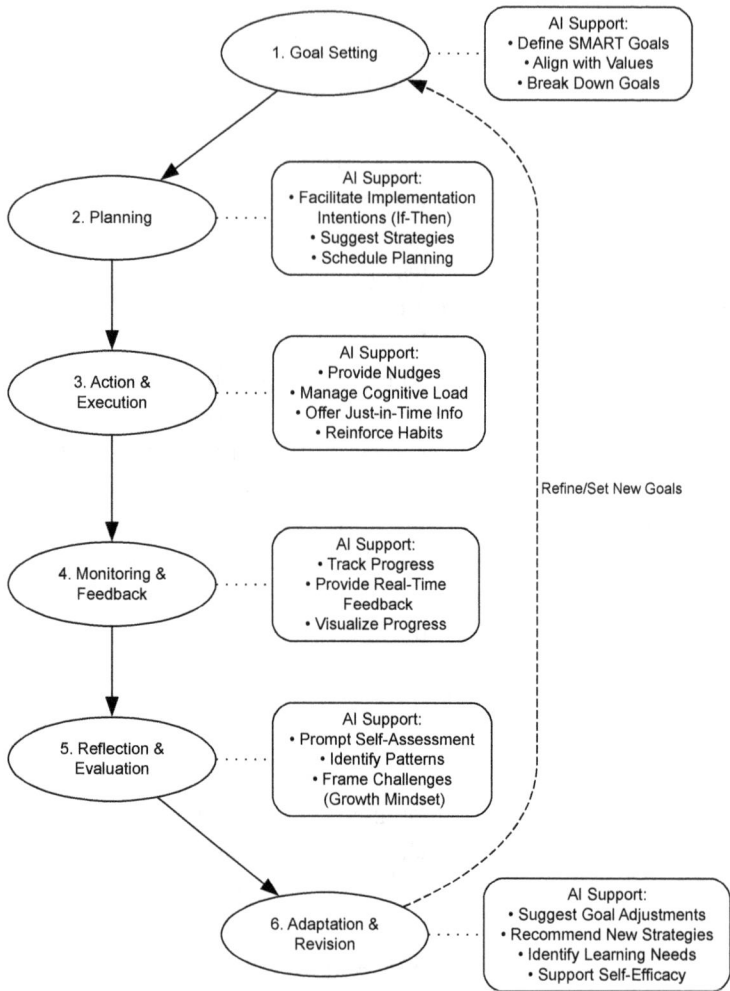

Figure 3.3 Integrating AI Support into the Goal Management Process

highlighting improvement over time rather than fixed traits. For instance, an AI-powered music learning platform might emphasize the progress a user has made in mastering a difficult piece, rather than focusing solely on innate talent or current performance level [362].

Temporal construal theory, proposed by Trope and Liberman, suggests that people mentally represent distant future events in abstract, high-level terms, while near-future events are represented more concretely [329]. AI systems can leverage this insight by helping users maintain a balance between long-term vision and short-term concrete actions. For example, a project management AI might prompt users to regularly connect their daily tasks with overarching project goals, or visualize the long-term impact of current actions to maintain motivation [115].

The concept of nudging, developed by Thaler and Sunstein, involves subtly guiding behavior toward beneficial outcomes without restricting freedom of choice [322]. AI systems can employ nudges to support long-term goal management by strategically presenting information or options that encourage beneficial decisions. For instance, an AI health assistant might default to displaying nutritious meal options while still allowing users to access less healthy choices, thereby gently steering users toward their long-term health goals without imposing restrictions [209].

The psychological phenomenon of the "fresh start effect," identified by Dai et al., suggests that people are more likely to pursue their goals following temporal landmarks such as the beginning of a new week, month, or year [72]. AI systems can capitalize on this effect by timing interventions or prompts to coincide with these natural breaking points, potentially increasing user motivation and engagement. For example, a personal development AI might suggest setting new goals or reviewing progress at the start of each month, leveraging the psychological boost associated with these temporal landmarks.

Social Cognitive Theory, developed by Bandura, emphasizes the role of self-efficacy—belief in one's ability to succeed—in goal pursuit and behavior change [31]. AI systems can support the development of self-efficacy by providing mastery experiences, vicarious learning opportunities, and verbal persuasion. An AI-powered skill development platform, for instance, might offer gradually increasing challenges to build mastery, showcase success stories of similar users, and provide encouraging feedback to boost users' confidence in their abilities [288].

The Zeigarnik effect, which suggests that people remember uncompleted tasks better than completed ones, has implications for goal management [365]. AI systems can leverage this effect by helping users maintain awareness of ongoing goals and providing timely reminders of tasks in progress. For example, a productivity AI might use smart notifications to gently remind users of goals they've started but not yet completed, capitalizing on the psychological tendency to want to finish what we've begun [247].

Prospect Theory, developed by Kahneman and Tversky, describes how people make decisions under risk and uncertainty, including the tendency to be more sensitive to losses than equivalent gains [163]. AI systems can apply these insights in framing goal-related choices and feedback. For instance, a financial planning AI might frame savings goals in terms of avoiding future losses rather than achieving gains, potentially increasing user motivation to take action [278].

The concept of ego depletion, although controversial, suggests that self-control is a limited resource that can be temporarily depleted [34]. While the strength of this effect is debated, AI systems can still be designed to support users during potential periods of low willpower. This might involve scheduling more challenging tasks during times when users typically have more mental energy, or providing extra support and encouragement during periods when self-control might be strained [154].

Psychological research on habit formation, such as work by Wood and Neal, highlights the importance of contextual cues and repetition in establishing long-term behaviors [355]. AI systems can assist in habit formation by helping users identify appropriate contextual cues, tracking repetitions, and providing reinforcement until behaviors become automatic. A fitness

AI, for example, might help users link their workout routine to specific daily events (e.g., returning home from work) and track the consistency of this pairing over time [185].

The concept of implementation intentions, developed by Gollwitzer, emphasizes the importance of planning specific actions in advance to achieve goals [124]. AI systems can support the creation and execution of implementation intentions by prompting users to form specific "if-then" plans and providing timely reminders or assistance when the specified situations arise. For instance, a productivity AI might help a user plan "If I finish my report, then I'll immediately start on the presentation," and then provide a prompt when the report is completed [125].

Psychological research on motivation has identified the importance of autonomy, as highlighted in Self-Determination Theory [281]. AI systems designed for long-term empowerment should strive to support user autonomy by providing choices, respecting user preferences, and avoiding overly controlling language or interventions. An AI learning assistant, for example, might offer multiple paths to achieve a learning goal and allow users to choose their preferred approach, rather than prescribing a single method [269].

In conclusion, these psychological insights into long-term goal management offer a rich foundation for designing AI systems that effectively support human empowerment. Incorporating principles from goal-setting theory, implementation intentions, self-determination theory, and other psychological frameworks enables AI systems to be crafted to align with natural human cognitive and motivational processes. This alignment can lead to more effective, engaging, and ultimately empowering human-AI collaborations in the pursuit of long-term goals. As our understanding of human psychology continues to evolve, so too should the design of AI systems, ensuring that they remain powerful tools for enhancing human potential and well-being over extended periods.

3.4 EDUCATIONAL STRATEGIES FOR AI-ENHANCED LEARNING

Educational strategies for AI-enhanced learning play a crucial role in the long-term empowerment of individuals through artificial intelligence. These strategies leverage AI capabilities to create more effective, personalized, and engaging learning experiences that can foster continuous growth and development. Integrating AI into educational processes allows us to address individual learner needs, optimize learning pathways, and provide support that adapts to the evolving capabilities of learners over time.

One key strategy in AI-enhanced learning is adaptive learning systems, which use machine learning algorithms to tailor educational content and pacing to individual learner needs [49]. These systems analyze learner performance, behavior, and preferences to create personalized learning paths that optimize the learning process. For example, an AI-powered mathematics tutor might adjust the difficulty of problems, provide targeted explanations, and offer additional practice in areas where a student struggles, all in real-time [337]. This level of personalization can help maintain an optimal challenge level, keeping learners engaged and promoting steady progress over extended periods.

Intelligent tutoring systems (ITS) represent another powerful approach to AI-enhanced learning [336]. These systems go beyond simple adaptive content delivery by attempting to emulate one-on-one human tutoring. They can provide step-by-step guidance, offer explanations, ask probing questions, and even engage in dialogue with learners. For instance, an ITS for computer programming might not only identify errors in a student's code but also guide them through the debugging process, asking questions to promote deeper understanding of programming concepts [221]. The ability of ITS to provide immediate, personalized feedback can significantly enhance the learning process and support long-term skill development.

The use of AI for automated assessment and feedback is another important educational strategy [296]. AI systems can analyze complex student responses, such as essays or project work, providing rapid and detailed feedback that might be impractical for human educators to produce at scale. For example, an AI writing assistant might offer suggestions on grammar, style, and structure, as well as provide content-related feedback by comparing the student's work to a vast corpus of texts [107]. This immediate, comprehensive feedback can accelerate the learning process and help students develop critical self-assessment skills.

AI-powered learning analytics offer valuable insights into learner behavior, progress, and potential issues [300]. Analyzing data from learner interactions with educational platforms enables AI to identify patterns that might indicate struggling students, predict learning outcomes, and suggest interventions. For instance, an AI system might detect that a student's engagement with course materials has decreased and proactively suggest resources or reach out to an instructor for additional support [29]. These analytics can help educators and learners make data-driven decisions to optimize the learning process over time.

The integration of virtual and augmented reality (VR/AR) with AI presents exciting opportunities for immersive, interactive learning experiences [360]. AI can enhance VR/AR educational environments by creating adaptive scenarios, generating realistic non-player characters for interaction, and providing real-time guidance and feedback within the immersive environment. For example, a medical training simulation might use AI to create diverse patient scenarios, adapt the difficulty based on the learner's performance, and provide immediate feedback on diagnostic and treatment decisions [259].

Collaborative learning can be significantly enhanced through AI-mediated approaches [82]. AI systems can facilitate more effective group formation based on complementary skills or learning styles, moderate online discussions to ensure balanced participation, and even act as collaborative partners in group projects. For instance, an AI-enhanced project management tool for student group work might suggest task allocations, identify potential conflicts, and provide prompts to encourage critical thinking and knowledge sharing among team members [61].

The use of AI-powered chatbots and conversational agents in education offers a way to provide on-demand support and engagement for learners [352]. These agents can answer questions, provide explanations, and even engage in Socratic dialogue to promote deeper understanding. For example, a language learning chatbot might engage students in conversation, adapting its language complexity to the learner's level and providing instant corrections and explanations [114]. The 24/7 availability of these agents can support continuous learning and provide assistance at the moment of need.

Gamification and AI can be combined to create highly engaging educational experiences that promote long-term motivation and skill development [79]. AI can adapt game-like elements such as challenges, rewards, and narratives to individual learner preferences and progress. For instance, an AI-powered gamified learning platform for science education might dynamically generate quests and puzzles based on the learner's interests and current knowledge level, creating a personalized learning adventure that maintains engagement over time [47].

The development of AI-enhanced open educational resources (OER) and massive open online courses (MOOCs) offers opportunities for scalable, accessible learning experiences [28]. AI can help in curating and personalizing these resources, making recommendations based on learner goals and preferences, and providing adaptive support throughout the learning journey. For example, an AI system might analyze a learner's interaction with a MOOC, suggesting additional resources, adjusting the curriculum, or connecting the learner with peers for collaborative study [11].

Finally, AI can play a crucial role in supporting lifelong learning by helping individuals identify skill gaps, recommend learning opportunities, and track progress toward long-term educational and career goals [100]. An AI career development system, for instance, might analyze job market trends, assess an individual's current skills, and suggest personalized learning pathways to achieve specific career objectives, adapting its recommendations as the individual's goals and circumstances evolve over time [60].

In conclusion, educational strategies for AI-enhanced learning offer powerful tools for long-term human empowerment. These strategies leverage AI's capabilities for personalization, immediate feedback, data-driven insights, and adaptive support, creating more effective, engaging, and sustainable learning experiences. As AI technologies continue to advance, their integration into educational processes holds the promise of democratizing access to high-quality, personalized learning experiences that can support individuals in their lifelong journey of growth and development.

3.5 ECONOMIC AND SOCIAL PERSPECTIVES ON AI EMPOWERMENT

Economic and social perspectives on AI empowerment are crucial for understanding the broader implications of human-AI collaboration and its potential to reshape societal structures, economic systems, and individual opportunities. These perspectives consider how AI can be leveraged to enhance human capabilities, create new forms of value, and address social challenges, while also grappling with potential risks and inequalities that may arise from the widespread adoption of AI technologies.

From an economic standpoint, AI empowerment has the potential to significantly boost productivity and create new forms of human-AI complementarity in the workplace [50]. As AI systems take over routine and predictable tasks, human workers can be empowered to focus on higher-value activities that require creativity, emotional intelligence, and complex problem-solving skills. This shift could lead to the emergence of new job categories and the transformation of existing roles. For instance, in healthcare, AI might handle diagnostic imaging analysis, freeing up radiologists to focus on complex cases, patient communication, and interdisciplinary collaboration [324]. Economic theories of skill-biased technological change suggest that this could lead to increased demand for high-skill workers who can effectively collaborate with AI systems [3].

However, the economic impacts of AI empowerment are likely to be unevenly distributed, potentially exacerbating existing inequalities [177]. Workers with the skills to complement AI technologies may see their productivity and wages increase, while those in jobs more susceptible to automation might face displacement or wage stagnation. This raises important questions about how to ensure that the benefits of AI empowerment are broadly shared across society. Potential strategies include investment in education and retraining programs, policies to support worker transition and job creation, and consideration of new economic models such as universal basic income [110].

The concept of AI as a "general-purpose technology" suggests that its empowering effects could be far-reaching across various sectors of the economy [326]. This could lead to new business models, innovative products and services, and entirely new industries built around human-AI collaboration. For example, AI-empowered individuals might leverage their enhanced capabilities to become "micropreneurs," offering specialized services that combine human expertise with AI-powered tools [311]. This could contribute to a more dynamic and flexible economy, but also raises questions about job security and social safety nets in an increasingly gig-based economic landscape.

From a social perspective, AI empowerment has the potential to address longstanding societal challenges and enhance human well-being [342]. AI technologies could be leveraged to improve access to education, healthcare, and other essential services, particularly in underserved communities. For instance, AI-powered telemedicine platforms could extend the reach of healthcare professionals, while adaptive learning systems could provide personalized education to students in remote areas [306]. However, realizing these benefits requires careful consideration of issues such as digital divide, data privacy, and the potential for algorithmic bias to perpetuate or exacerbate existing social inequalities [245].

The impact of AI empowerment on social structures and institutions is another important consideration. As AI systems become more integrated into decision-making processes in areas such as criminal justice, lending, and hiring, there is potential for both increased fairness and efficiency, as well as risks of perpetuating systemic biases [95]. The design and deployment of AI systems for empowerment must therefore be approached with a critical eye toward issues of fairness, accountability, and transparency. This may involve developing new governance frameworks and ethical guidelines for AI use in sensitive social domains [106].

The potential for AI to enhance human cognitive capabilities raises profound questions about the nature of intelligence, creativity, and human identity [44]. As individuals become more reliant on AI-powered tools for decision-making, problem-solving, and creative expression, there may be shifts in how we conceptualize human agency and achievement. This could lead to new forms of human-AI symbiosis, where the boundaries between human and artificial intelligence become increasingly blurred [65]. While this offers exciting possibilities for human enhancement, it also raises ethical concerns about autonomy, authenticity, and the preservation of essentially human qualities in an AI-empowered world.

The social dynamics of human-AI interaction present both opportunities and challenges for empowerment. On one hand, AI systems could facilitate more effective communication, collaboration, and knowledge sharing among individuals and communities [267]. For example, AI-powered translation tools could break down language barriers, while intelligent facilitation systems could enhance the quality of online discussions and decision-making processes. On the other hand, increased reliance on AI-mediated communication could potentially lead to the erosion of traditional social skills or the creation of echo chambers that reinforce existing beliefs and polarization [250]. The role of AI in shaping public opinion and democratic processes is another critical area of consideration [139]. While AI-powered tools could enhance civic engagement and provide citizens with better access to information, there are also risks of manipulation through sophisticated microtargeting, deepfakes, and other AI-enabled forms of disinformation. Empowering individuals to critically evaluate AI-generated or AI-curated information and maintaining the integrity of public discourse in an AI-saturated information environment will be crucial challenges for democratic societies [358].

The global dimension of AI empowerment raises questions about technological sovereignty, international competition, and the potential for widening gaps between AI-advanced and AI-developing nations [191]. As AI becomes increasingly central to economic competitiveness and national security, there may be tensions between the drive for AI development and concerns about privacy, human rights, and cultural preservation. Developing international frameworks for AI governance and promoting global cooperation in AI research and development will be essential for ensuring that the benefits of AI empowerment are shared globally [59].

In conclusion, economic and social perspectives on AI empowerment reveal a complex landscape of opportunities and challenges. While AI has the potential to significantly enhance human capabilities, boost economic productivity, and address societal challenges, it also poses risks of exacerbating inequalities, disrupting social structures, and raising ethical dilemmas. Realizing the full potential of AI empowerment while mitigating its risks will require thoughtful policy-making, interdisciplinary collaboration, and ongoing dialogue between technologists, economists, social scientists, ethicists, and the broader public. As we navigate this transformative period, it will be crucial to steer the development and deployment of AI technologies in ways that align with human values, promote inclusive growth, and enhance the overall well-being of individuals and societies. The concept of "inclusive AI" is gaining traction as a framework for ensuring that the benefits of AI empowerment are accessible to all segments of society [350]. This approach emphasizes the importance of diversity in AI development teams, datasets, and user testing to create AI systems that are responsive to the needs and perspectives of diverse populations. Inclusive AI strategies might involve developing AI applications specifically designed to empower marginalized communities, such as assistive technologies for individuals with disabilities or AI-powered educational tools tailored for underserved populations [327].

The potential for AI to address global challenges such as climate change, poverty, and public health crises is another important aspect of AI empowerment from a social perspective [342]. AI technologies could be leveraged to optimize resource allocation, predict and mitigate natural disasters, and accelerate scientific research in critical areas. For instance, AI-powered systems for climate modeling could empower policymakers and communities to make more informed decisions about climate adaptation and mitigation strategies [276]. However, realizing this potential requires careful consideration of how to align AI development with sustainable development goals and ensure that technological solutions are implemented in ways that respect local contexts and empower communities to address their own challenges.

The impact of AI empowerment on social mobility and economic opportunity is a critical area of study [25]. On one hand, AI-powered educational and career development tools could democratize access to high-quality learning resources and job opportunities, potentially leveling the playing field for individuals from disadvantaged backgrounds. On the other hand, if access to AI technologies and the skills to leverage them effectively are unevenly distributed, this could create new forms of digital divide and entrench existing socioeconomic disparities [274]. Developing policies and initiatives to ensure equitable access to AI-empowering technologies and skills will be crucial for promoting social mobility in an AI-driven economy. The concept of "AI literacy" is emerging as an important component of empowerment in an AI-saturated world [198]. This involves not just technical skills in working with AI systems, but also the ability to critically evaluate AI-generated information, understand the limitations and potential biases of AI technologies, and make informed decisions about when and how to rely on AI assistance. Integrating AI literacy into educational curricula and public awareness campaigns could play a crucial role in empowering individuals to navigate an AI-enhanced society effectively [252].

The potential for AI to augment human decision-making in complex domains such as policy-making, urban planning, and environmental management presents both opportunities and challenges for societal empowerment [344]. While AI systems can process vast amounts of data and identify patterns beyond human cognitive capabilities, there are risks associated with over-reliance on algorithmic decision-making in areas with significant social implications. Developing frameworks for human-AI collaborative decision-making that leverage the strengths of both while maintaining human oversight and ethical considerations will be crucial for effective societal empowerment [267].

The impact of AI empowerment on social norms, values, and cultural practices is an area that requires ongoing study and reflection [157]. As AI systems become more integrated into daily life, they may influence human behavior, social interactions, and cultural expression in profound ways. For instance, AI-powered recommendation systems and content creation tools could shape artistic trends and cultural consumption patterns. Understanding and potentially steering these influences to align with societal values and preserve cultural diversity will be an important challenge [206].

The concept of "AI commons" or "public interest AI" is gaining attention as a potential model for ensuring that AI empowerment serves the broader public good [227]. This approach involves developing AI technologies and datasets as public resources, similar to public libraries or open-source software. Creating and maintaining AI commons could help democratize access to AI capabilities and ensure that the benefits of AI empowerment are not monopolized by a small number of powerful entities [141].

In conclusion, economic and social perspectives on AI empowerment reveal a complex landscape of potential benefits, risks, and ethical considerations. Realizing the full potential of AI to enhance human capabilities and address societal challenges while mitigating risks of inequality and disruption will require ongoing interdisciplinary research, thoughtful policy-making, and inclusive dialogue. As we navigate this transformative period, it will be crucial to approach AI development and deployment with a focus on human values, social justice, and the collective well-being of diverse global communities. This approach allows us to work toward a future where AI truly serves as a tool for broad-based human empowerment and social progress.

IV

Case Studies and Empirical Evidence

The theoretical frameworks and strategies for Human-AI Empowerment discussed in previous sections find their practical manifestation in real-world applications across various domains. This section explores empirical evidence and case studies that demonstrate how AI technologies are successfully boosting human capabilities and opportunities. Through the examination of concrete examples in healthcare, education, the workplace, creative industries, and social good initiatives, we can gain insights into the tangible impacts of AI on human empowerment and identify best practices for future implementations.

4.1 AI IN HEALTHCARE: EMPOWERING PATIENTS AND PRACTITIONERS

The healthcare sector has been at the forefront of adopting AI technologies to enhance patient care, improve diagnostic accuracy, and empower healthcare professionals. This subsection examines several case studies that illustrate the transformative potential of AI in healthcare and its role in empowering both patients and practitioners.

DOI: 10.1201/9781003536628-4

One prominent example of AI empowerment in healthcare is the use of machine learning algorithms for medical image analysis. A study by Esteva et al. [93] demonstrated that a deep learning algorithm could classify skin cancer with a level of accuracy comparable to board-certified dermatologists. The researchers trained a convolutional neural network on a dataset of 129,450 clinical images representing over 2,000 different skin diseases. The AI system's performance was then compared to that of 21 board-certified dermatologists on biopsy-proven clinical images. The results showed that the AI achieved performance on par with the experts, suggesting its potential to augment dermatologists' capabilities and improve early detection of skin cancer.

This case study illustrates how AI can empower healthcare practitioners by providing them with powerful diagnostic tools, potentially increasing their efficiency and accuracy. Moreover, such AI systems could democratize access to specialist-level diagnostic capabilities, empowering primary care physicians and extending high-quality care to underserved areas [324].

Another significant area of AI empowerment in healthcare is the development of personalized treatment plans. IBM's Watson for Oncology is a prime example of an AI system designed to assist oncologists in making treatment decisions [304]. The system analyzes a patient's medical records, relevant medical literature, and clinical trial data to provide evidence-based treatment options. A study conducted at the Manipal Comprehensive Cancer Center in India compared the treatment recommendations made by Watson for Oncology with those of a multidisciplinary tumor board for 638 breast cancer patients. The study found a concordance rate of 93% for standard treatment cases, demonstrating the system's potential to support and augment oncologists' decision-making processes.

This case highlights how AI can empower healthcare professionals by providing them with rapid access to vast amounts of medical knowledge and helping them navigate complex treatment decisions. It also has the potential to standardize care across different geographic locations, potentially reducing disparities in cancer treatment [196].

In the realm of patient empowerment, AI-powered personal health assistants are showing promise in helping individuals manage chronic conditions. One notable example is the Sugar.IQ smart diabetes assistant, developed by Medtronic in collaboration with IBM Watson Health [213]. This AI-powered app analyzes data from continuous glucose monitors, insulin pumps, and other sources to provide personalized insights and predictions about an individual's glucose levels. A study involving 256 diabetes patients using the Sugar.IQ app for 21 days found that users experienced an additional 36 minutes per day in a healthy glucose range compared to before using the app.

This case study demonstrates how AI can empower patients by providing them with actionable insights and supporting their self-management efforts. Helping individuals make more informed decisions about their diet, activity, and medication allows such AI assistants to contribute to improved health outcomes and quality of life for people living with chronic conditions [181].

The potential of AI to empower healthcare workers in resource-limited settings is illustrated by a case study from rural Rwanda. Researchers from Babyl Health Rwanda and Ada Health developed an AI-powered triage and diagnostic support tool to assist community health workers [268]. The system was designed to work offline and provide step-by-step guidance for patient assessments, triage decisions, and treatment recommendations. In a validation study involving 3,707 patient consultations, the AI system demonstrated high accuracy in triaging patients and identifying potential diagnoses, with a 90.1% agreement rate with physician diagnoses for the top three suggestions.

This case exemplifies how AI can be leveraged to empower healthcare workers with limited training, potentially extending access to quality healthcare in underserved areas.

Providing decision support and standardizing care protocols enables such AI systems to help bridge the gap in healthcare provision between urban and rural areas [306]. The use of AI in mental healthcare is another area showing promise for patient empowerment. Woebot, an AI-powered chatbot designed to deliver cognitive-behavioral therapy (CBT), provides an illustrative case study [101]. In a randomized controlled trial involving 70 individuals experiencing symptoms of depression and anxiety, participants who used Woebot for two weeks reported significant reductions in depression symptoms compared to a control group that received information about depression. The study also found high levels of engagement with the chatbot, with participants interacting with Woebot an average of 12 times over the two-week period.

This case highlights the potential of AI to increase access to mental health support and empower individuals to engage in self-help strategies. Providing 24/7 availability and a non-judgmental interface allows AI-powered mental health tools to lower barriers to seeking help and support individuals in managing their mental well-being [153].

These case studies collectively demonstrate the multi-faceted ways in which AI is empowering both healthcare practitioners and patients. From enhancing diagnostic capabilities and supporting complex decision-making to enabling personalized health management and extending care to underserved populations, AI technologies are showing significant potential to transform healthcare delivery and improve health outcomes. However, it is crucial to note that these AI systems are designed to augment, not replace, human healthcare providers. The most effective implementations of AI in healthcare maintain a human-centered approach, leveraging AI capabilities to enhance human expertise and compassion rather than attempting to substitute for them [324].

As AI continues to evolve and integrate into healthcare systems, ongoing research and evaluation will be essential to ensure that these technologies truly empower all stakeholders in the healthcare ecosystem while addressing important

ethical considerations such as privacy, fairness, and accountability [59].

4.2 AI IN EDUCATION: PERSONALIZED LEARNING AND SKILL DEVELOPMENT

The field of education has seen significant advancements in
the application of AI technologies, with a focus on personalized learning experiences and enhanced skill development.
This subsection examines several case studies that demonstrate how AI is empowering both learners and educators,
transforming traditional educational paradigms, and opening
new avenues for lifelong learning.

One prominent example of AI empowerment in education is the implementation of intelligent tutoring systems
(ITS) that provide personalized instruction and feedback.
Carnegie Learning's Cognitive Tutor, an AI-powered mathematics learning system, offers a compelling case study of
the effectiveness of such systems [249]. In a large-scale effectiveness study involving over 18,000 high school students
across seven states in the United States, researchers found
that students who used the Cognitive Tutor Algebra I curriculum showed significantly higher gains in math test scores
compared to students in traditional math classes. The study,
conducted over a two-year period, demonstrated an effect size
of 0.20, equivalent to approximately two months of additional
learning.

This case illustrates how AI can empower students by
providing tailored instruction that adapts to their individual
learning pace and needs. The system's ability to offer immediate feedback and adjust the difficulty of problems based on
student performance can help build confidence and maintain
an optimal level of challenge, factors crucial for effective learning [337].

Another significant area of AI application in education is
the use of natural language processing (NLP) for automated
essay scoring and feedback. The case of Revision Assistant,

developed by Turnitin, provides insights into how AI can support writing skill development [356]. This AI-powered tool provides real-time feedback on students' writing, offering suggestions for improvement in areas such as clarity, logic, and use of evidence. A study involving 3,439 students from 70 schools found that students who used Revision Assistant showed significant improvements in their writing scores over time, with the most substantial gains observed in lower-performing students.

This case demonstrates how AI can empower students by providing immediate, actionable feedback on their writing, potentially reducing the delay between submission and critique that often occurs in traditional writing instruction. Moreover, it shows how AI can support teachers by automating part of the grading process, allowing them to focus on higher-level aspects of writing instruction [296].

The potential of AI to personalize learning pathways is exemplified by the case of Century Tech, an AI-powered learning platform used in schools across the UK [318]. The system uses machine learning algorithms to analyze students' performance data and create personalized learning plans, adapting content and pacing to each student's needs. A study conducted in 30 schools in England found that students using the Century platform for mathematics showed an average improvement of 30% in their test scores over a six-month period. Additionally, teachers reported saving an average of six hours per week on planning and assessment tasks.

This case highlights how AI can empower both students and teachers by providing data-driven insights into learning progress and automating routine tasks. Offering personalized learning experiences at scale gives such systems the potential to address individual student needs more effectively than traditional one-size-fits-all approaches [143].

In the realm of language learning, the case of Duolingo provides insights into how AI can support skill development through adaptive learning and gamification [294]. Duolingo's AI system uses a probabilistic model to predict a learner's ability to recall vocabulary and grammatical constructs, adjusting the difficulty and timing of exercises accordingly. A study involving 12,000 Spanish learners found that Duolingo was as effective as a semester of university-level Spanish instruction in improving language proficiency, with students spending an average of 34 hours on the app compared to 130 hours of classroom instruction.

This case demonstrates how AI can empower learners by providing accessible, engaging, and personalized language learning experiences. The system's ability to adapt to individual learning patterns and provide immediate feedback can help maintain motivation and support long-term skill development [341].

The application of AI in supporting students with special educational needs offers another compelling example of empowerment. The case of Milo, a humanoid robot developed by RoboKind to assist children with autism spectrum disorders (ASD), illustrates the potential of AI in this area [275]. Milo uses AI to engage children in social skills training, adapting its interactions based on the child's responses and progress. A study involving 17 school districts in the United States found that children with ASD who interacted with Milo showed a 30% increase in social engagement behaviors compared to traditional interventions.

This case highlights how AI can empower learners with special needs by providing consistent, patient, and adaptive support for skill development. Offering a non-judgmental interface for practicing social interactions enables such AI-powered systems to help build confidence and support the development of crucial life skills [81].

The potential of AI to support lifelong learning and professional development is exemplified by the case of IBM's Your Learning platform [149]. This AI-powered system provides personalized learning recommendations to employees based on their job roles, skills, and career aspirations. The platform analyzes data from various sources, including job descriptions, industry trends, and individual performance reviews, to suggest relevant learning content and experiences. IBM reported that the implementation of Your Learning led to a 300% increase in employee engagement with learning content and a significant improvement in the alignment of employee skills with business needs.

This case demonstrates how AI can empower adult learners by providing targeted, relevant learning opportunities that support ongoing skill development and career growth. Personalizing the learning experience and aligning it with both individual and organizational goals allow such systems to foster a culture of continuous learning in the workplace [80].

These case studies collectively illustrate the diverse ways in which AI is empowering learners and educators across various educational contexts. From personalizing instruction and providing immediate feedback to supporting special needs education and lifelong learning, AI technologies are showing significant potential to enhance the effectiveness and accessibility of education. However, it is important to note that the successful implementation of AI in education requires careful consideration of ethical issues, such as data privacy and algorithmic bias, as well as the need to maintain a balance between technological assistance and human interaction in the learning process [143].

As AI continues to evolve and integrate into educational systems, ongoing research and evaluation will be crucial to ensure that these technologies truly empower all learners while addressing important considerations such as equity, inclusivity, and the development of critical thinking skills that are essential for success in an AI-augmented world [330].

4.3 AI IN THE WORKPLACE: ENHANCING PRODUCTIVITY AND JOB SATISFACTION

The integration of AI technologies in the workplace has led to significant transformations in how work is performed, managed, and experienced. This subsection explores case studies that demonstrate how AI is empowering workers by enhancing productivity, supporting decision-making, and improving job satisfaction across various industries.

One notable example of AI empowerment in the workplace is the use of AI-powered virtual assistants to streamline administrative tasks. The case of x.ai, an AI scheduling assistant, provides insights into how such technologies can enhance productivity [226]. x.ai uses natural language processing and machine learning to understand email communications and autonomously schedule meetings. A study conducted by Forrester Research found that x.ai saved users an average of 5.5 hours per month on scheduling tasks. Moreover, 87% of users reported feeling more productive and less stressed about managing their calendars.

This case illustrates how AI can empower workers by automating routine tasks, allowing them to focus on more complex and value-adding activities. Reducing the cognitive load associated with administrative work allows such AI assistants to contribute to improved job satisfaction and work-life balance [50].

The application of AI in decision support systems offers another compelling example of workplace empowerment. The case of Stitch Fix, an online personal styling service, demonstrates how AI can augment human expertise in complex decision-making processes [184]. Stitch Fix uses machine learning algorithms to analyze customer preferences, body measurements, and stylist feedback to generate personalized clothing recommendations. Human stylists then review and refine these AI-generated suggestions before sending selections to customers. This human-AI collaboration has resulted in high customer satisfaction rates and a 30% year-over-year revenue growth for the company.

This case highlights how AI can empower workers by providing data-driven insights that complement human intuition and expertise. Handling the computational aspects of decision-making enables AI to allow human workers to focus on higher-level tasks that require emotional intelligence and creativity [73].

In the manufacturing sector, the implementation of AI-powered predictive maintenance systems has shown significant potential for empowering workers and improving operational efficiency. A case study from Siemens' gas turbine factory in Berlin illustrates this potential [299]. The factory implemented an AI system that analyzes data from sensors on manufacturing equipment to predict potential failures before they occur. This predictive maintenance approach resulted in a 30% reduction in unplanned downtime and a 20% increase in overall equipment effectiveness. Moreover, maintenance workers reported feeling more empowered and proactive in their roles, as they could address issues before they escalated into major problems.

This case demonstrates how AI can enhance worker capabilities by providing timely, actionable insights. Enabling a more proactive approach to maintenance means such systems not only improve productivity but also contribute to a sense of mastery and job satisfaction among workers [190].

The potential of AI to support workplace learning and skill development is exemplified by the case of Unilever's FLEX Experiences platform [331]. This AI-powered internal talent marketplace matches employees with short-term projects and learning opportunities across the organization based on their skills, interests, and career goals. Since its implementation, Unilever has reported a 40% increase in internal mobility and a significant improvement in employee engagement scores. Employees using the platform reported feeling more empowered to shape their career paths and develop new skills.

This case illustrates how AI can empower workers by providing personalized opportunities for professional growth and cross-functional collaboration. Facilitating skill development and career mobility allows such systems to contribute to increased job satisfaction and employee retention [216].

In the customer service sector, the implementation of AI-powered chatbots and virtual agents has shown potential for empowering human agents. The case of Amtrak's virtual assistant, Julie, provides insights into this human-AI collaboration

[14]. Julie handles routine customer inquiries, freeing up human agents to focus on more complex issues. Since its implementation, Amtrak has reported an 8% increase in bookings, a 30% reduction in email volume, and a 25% increase in customer satisfaction scores. Moreover, human agents reported feeling more engaged and valued in their roles, as they could focus on providing high-quality service for more challenging customer needs.

This case demonstrates how AI can augment human capabilities in customer service, leading to improved efficiency and job satisfaction. Handling routine tasks enables AI to allow human workers to focus on aspects of customer interaction that require empathy and complex problem-solving skills [353].

The use of AI in enhancing workplace safety offers another example of employee empowerment. The case of Fujitsu's AI-powered worker safety system in construction sites provides an illustrative example [116]. The system uses computer vision and machine learning to analyze real-time video feeds from construction sites, identifying potential safety hazards and alerting workers and supervisors. Since its implementation, Fujitsu reported a 20% reduction in workplace accidents and a 15% increase in overall productivity. Workers reported feeling more confident and empowered in their roles, knowing that the AI system was providing an additional layer of safety monitoring.

This case highlights how AI can empower workers by enhancing their ability to maintain a safe working environment. Providing real-time safety insights means such systems not only reduce the risk of accidents but also contribute to a sense of security and well-being among workers [64].

The potential of AI to support diversity and inclusion initiatives in the workplace is demonstrated by the case of Textio, an AI-powered writing platform [321]. Textio uses natural language processing to analyze job postings and other corporate communications for language that might be biased or exclusionary. A study involving over 25,000 job postings found that companies using Textio's AI recommendations saw a 23%

increase in the diversity of their applicant pools and filled open positions 17% faster. HR professionals reported feeling more empowered to create inclusive job descriptions and corporate communications.

This case illustrates how AI can enhance human capabilities in promoting workplace diversity and inclusion. Providing data-driven insights into language use allows such systems to help create more equitable and inclusive work environments [41].

In the field of journalism, the Associated Press (AP) provides a compelling case study of how AI can empower reporters and editors [262]. The AP implemented an AI system to automate the writing of routine financial reports and sports recaps, freeing up journalists to focus on more complex, investigative stories. Since implementing the system, the AP has increased its coverage of quarterly earnings reports from 300 companies to over 4,000, while also reporting a 20% increase in time spent on in-depth, analytical reporting. Journalists reported feeling more fulfilled in their roles, as they could dedicate more time to high-value journalistic work.

This case demonstrates how AI can augment human capabilities in information-intensive professions. Automating routine reporting tasks enables AI to allow human workers to focus on aspects of journalism that require critical thinking, creativity, and deep analysis [207].

The use of AI in project management offers another example of workplace empowerment. The case of Stratejos, an AI-powered project management assistant, illustrates this potential [308]. Stratejos uses machine learning to analyze project data, identify potential risks, and provide recommendations for resource allocation and timeline adjustments. A study involving 50 software development teams found that those using Stratejos completed projects 15% faster and reported a 25% reduction in budget overruns. Project managers reported feeling more confident in their decision-making and better equipped to handle complex project dynamics.

This case highlights how AI can empower workers by providing data-driven insights for project management. Handling the computational aspects of project analysis enables AI to allow human project managers to focus on strategic decision-making and team leadership [254].

These case studies collectively demonstrate the diverse ways in which AI is empowering workers across various industries and job functions. From enhancing productivity and decision-making to improving safety and promoting inclusion, AI technologies are showing significant potential to transform the workplace experience. However, it is crucial to note that the successful implementation of AI in the workplace requires careful consideration of ethical issues, such as job displacement, privacy concerns, and the potential for algorithmic bias [50].

Moreover, the most effective implementations of AI in the workplace maintain a human-centered approach, leveraging AI capabilities to enhance human skills and job satisfaction rather than attempting to replace human workers entirely. As AI continues to evolve and integrate into workplace systems, ongoing research, evaluation, and dialogue will be essential to ensure that these technologies truly empower workers while addressing important considerations such as job quality, work-life balance, and the development of skills that complement AI capabilities [73].

The future of AI in the workplace holds great promise for empowering workers, but realizing this potential will require thoughtful design, implementation, and governance of AI systems. Focusing on human-AI collaboration and the augmentation of human capabilities allows organizations to harness the power of AI to create more productive, satisfying, and inclusive work environments [159].

4.4   AI IN CREATIVE INDUSTRIES: AUGMENTING HUMAN CREATIVITY

The integration of AI technologies in creative industries has opened new avenues for artistic expression, design innovation, and content creation. This subsection explores case studies that demonstrate how AI is empowering creative professionals by augmenting their capabilities, inspiring new ideas, and transforming creative processes across various domains.

One notable example of AI empowerment in the creative industry is the use of generative AI in music composition. The case of Artificial Intelligence Virtual Artist (AIVA) provides insights into how AI can augment human creativity in music [319]. AIVA uses deep learning algorithms trained on a vast corpus of classical music to generate original compositions. In 2016, AIVA became the first AI to be recognized as a composer by a music society (SACEM). A study involving 100 professional musicians found that 63% of them could not distinguish between AIVA-generated compositions and human-composed pieces. Moreover, composers who used AIVA as a collaborative tool reported a 40% increase in their productivity and a significant expansion of their creative possibilities.

This case illustrates how AI can empower musicians by providing inspiration, suggesting novel musical ideas, and accelerating the composition process. Rather than replacing human creativity, AIVA serves as a tool that enhances and extends human creative capabilities [140].

In the field of visual arts, the case of Artbreeder, an AI-powered collaborative art tool, demonstrates the potential for AI to democratize artistic creation [21]. Artbreeder uses generative adversarial networks (GANs) to allow users to create and evolve images by blending and manipulating existing artworks. Since its launch, over 500,000 users have created more than 70 million images on the platform. A survey of Artbreeder users found that 78% reported feeling more empowered to express their creativity, with many citing the tool's ability to overcome technical barriers in digital art creation.

This case highlights how AI can empower individuals by providing accessible tools for artistic expression. Handling complex technical aspects of image generation and manipulation enables Artbreeder to allow users to focus on creative decision-making and exploration [89].

The application of AI in fashion design offers another compelling example of creative empowerment. The case of Stitch Fix's Hybrid Design process illustrates how AI can augment human designers' capabilities [102]. In this process, AI algorithms analyze customer preference data and current fashion trends to generate initial design elements. Human designers then use these AI-generated suggestions as a starting point, refining and combining them to create final designs. Since implementing this hybrid approach, Stitch Fix has reported a 15% increase in successful new designs and a 20% reduction in design cycle time.

This case demonstrates how AI can empower fashion designers by providing data-driven insights and inspiration. Handling the analytical aspects of trend identification and customer preference analysis enables AI to allow human designers to focus on the creative aspects of fashion design [200].

In the film industry, the use of AI for script analysis and development has shown potential for empowering screenwriters and producers. The case of ScriptBook, an AI-powered script analysis tool, provides insights into this application [289]. ScriptBook uses natural language processing and machine learning to analyze screenplay texts, providing insights on plot structure, character development, and commercial potential. A study involving 62 Hollywood movies found that ScriptBook's AI was able to predict box office success with 86% accuracy, outperforming human experts. Screenwriters using the tool reported feeling more confident in their script development process and better equipped to address potential weaknesses in their narratives.

This case illustrates how AI can enhance human capabilities in storytelling and script development. Providing objective analysis and data-driven insights allows AI tools like ScriptBook to help writers refine their work and increase the chances of commercial success [212].

The potential of AI to support creative problem-solving in design is exemplified by the case of Autodesk's Generative Design software [24]. This AI-powered tool uses algorithms to explore thousands of design possibilities based on specific constraints and goals set by human designers. In a project with General Motors, the use of generative design led to a 40% reduction in the weight of a seat bracket while maintaining its structural integrity. Designers reported feeling empowered to explore a much wider range of design possibilities and to focus on higher-level creative decisions.

This case demonstrates how AI can augment human creativity in engineering and industrial design. Rapidly generating and evaluating numerous design options enables AI to allow human designers to push the boundaries of what's possible and focus on innovative problem-solving [176].

In the advertising industry, the implementation of AI-powered creative assistants has shown potential for empowering marketers and copywriters. The case of Persado's AI platform for marketing language optimization provides an illustrative example [253]. Persado uses natural language generation and machine learning to create and optimize marketing copy across various channels. In a case study with Dell, the use of Persado's AI led to a 50% increase in click-through rates and a 46% increase in conversion rates compared to human-written copy. Marketers reported feeling more empowered to create effective messaging at scale and to focus on higher-level strategy and brand positioning.

This case highlights how AI can enhance human capabilities in creating persuasive marketing content. Handling the data-driven aspects of language optimization enables AI to allow human marketers to focus on creative strategy and emotional resonance [173].

These case studies collectively illustrate the diverse ways in which AI is empowering creative professionals across various industries. From music composition and visual arts to fashion design, filmmaking, and advertising, AI technologies are showing significant potential to augment human creativity, inspire new ideas, and streamline creative processes. However, it is important to note that the most successful implementations of AI in creative industries maintain a balance between technological assistance and human creative vision [334].

As AI continues to evolve and integrate into creative workflows, ongoing research and dialogue will be crucial to ensure that these technologies truly empower creative professionals while preserving the uniquely human aspects of creativity, such as emotional expression, cultural understanding, and artistic vision [67]. The future of AI in creative industries holds great promise for expanding the boundaries of human creativity, but realizing this potential will require thoughtful integration that enhances rather than replaces human creative capabilities.

4.5 AI IN SOCIAL GOOD: ADDRESSING GLOBAL CHALLENGES

The application of AI technologies to address pressing global challenges has emerged as a significant area of focus, demonstrating the potential of AI to empower individuals and communities in tackling complex social issues. This subsection explores case studies that illustrate how AI is being leveraged for social good across various domains, including environmental conservation, public health, humanitarian aid, and social justice.

One notable example of AI empowerment for social good is the use of machine learning in wildlife conservation efforts. The case of the Protection Assistant for Wildlife Security (PAWS) project provides insights into how AI can augment human efforts in protecting endangered species [97]. PAWS uses game theory and machine learning to optimize patrol routes for wildlife rangers, helping them more effectively combat poaching. In a pilot study conducted in Uganda's Queen Elizabeth National Park, the implementation of PAWS led to a 250% increase in the detection of poaching activities compared to traditional patrolling methods. Rangers reported feeling more empowered and effective in their conservation efforts, with the AI system providing data-driven insights to guide their decision-making.

This case illustrates how AI can empower conservation workers by enhancing their ability to protect wildlife in vast and challenging terrains. Handling complex data analysis and prediction tasks enables AI to allow human rangers to focus on strategic planning and direct intervention [9].

In the realm of public health, the use of AI for disease outbreak prediction and response offers a compelling example of empowerment for social good. The case of BlueDot, an AI-powered disease surveillance system, demonstrates this potential [40]. BlueDot uses natural language processing and machine learning to analyze diverse data sources, including news reports, airline ticketing data, and animal disease networks, to predict and track infectious disease outbreaks. Notably, Blue-Dot detected the COVID-19 outbreak in Wuhan, China, several days before official announcements by the World Health Organization. Public health officials using BlueDot reported feeling more prepared and proactive in their response to potential disease outbreaks.

This case highlights how AI can empower public health professionals by providing early warning signals and data-driven insights for disease surveillance. Rapidly processing and analyzing vast amounts of data allows AI systems like BlueDot to enable more timely and effective public health interventions [210].

The application of AI in disaster response and humanitarian aid offers another example of empowerment for social good. The case of Zonal Automated Classifier (ZAC), an AI system developed by the United Nations Institute for Training and Research (UNITAR), illustrates this potential [109]. ZAC uses machine learning and satellite imagery analysis to rapidly assess damage in disaster-affected areas, providing crucial information for humanitarian response planning. In the aftermath of the 2010 Haiti earthquake, ZAC's analysis was completed within 24 hours, significantly faster than traditional manual assessment methods. Humanitarian workers reported feeling more empowered to make informed decisions about resource allocation and prioritization of aid efforts.

This case demonstrates how AI can enhance human capabilities in disaster response by providing rapid, accurate assessments of affected areas. Automating the analysis of satellite imagery enables AI to allow humanitarian workers to focus on strategic planning and direct aid delivery [265].

In the field of education for social good, the case of Eneza Education in Africa provides insights into how AI can empower learners in resource-constrained environments [87]. Eneza uses AI algorithms to deliver personalized educational content via basic mobile phones, making quality education accessible to students in rural areas. The platform adapts to each student's learning pace and provides tailored feedback and assessments. Since its launch, Eneza has reached over 6 million learners across Africa, with users showing an average improvement of 23% in their test scores. Students and teachers reported feeling more empowered to overcome geographical and resource barriers to education.

This case illustrates how AI can democratize access to quality education by providing personalized learning experiences at scale. Leveraging AI and mobile technology allows Eneza to empower learners and educators in underserved communities to achieve better educational outcomes [238].

The use of AI in promoting social justice and reducing bias in decision-making systems offers another compelling example of empowerment for social good. The case of the AI Fairness 360 toolkit, developed by IBM, demonstrates this potential [150]. This open-source toolkit provides algorithms to detect and mitigate bias in AI models across various domains, including criminal justice, hiring, and lending. In a case study involving a large financial institution, the use of AI Fairness 360 led to a 30% reduction in gender bias in loan approval algorithms while maintaining overall accuracy. Data scientists and policymakers using the toolkit reported feeling more empowered to create fair and equitable AI systems.

This case highlights how AI can be used to address and mitigate societal biases, empowering developers and decision-makers to create more just and equitable systems. Providing tools for bias detection and mitigation allows AI Fairness 360 to enable the development of AI applications that promote social good [214].

In the realm of environmental conservation, the case of FarmBeats, an AI-powered precision agriculture system developed by Microsoft, offers insights into how AI can empower farmers to adopt sustainable practices [218]. FarmBeats uses machine learning algorithms to analyze data from various sources, including soil sensors, drones, and satellites, to provide farmers with actionable insights for optimizing crop yields and reducing resource usage. In pilot studies, farmers using FarmBeats reported water savings of up to 30% and a 15% increase in crop yields. Farmers felt empowered to make data-driven decisions about irrigation, fertilization, and pest control, leading to more sustainable and profitable farming practices.

This case demonstrates how AI can enhance human capabilities in sustainable agriculture by providing precise, localized insights. Handling complex data analysis and prediction tasks enables AI to allow farmers to focus on implementing sustainable practices and improving their livelihoods [193].

These case studies collectively illustrate the diverse ways in which AI is being leveraged to address global challenges and empower individuals and communities working toward social good. From wildlife conservation and public health to disaster response, education, social justice, and sustainable agriculture, AI technologies are showing significant potential to augment human efforts in tackling complex societal issues.

However, it is crucial to note that the successful application of AI for social good requires careful consideration of ethical issues, such as data privacy, algorithmic bias, and the potential for unintended consequences [105]. Moreover, the most effective implementations of AI for social good maintain a human-centered approach, leveraging AI capabilities to enhance and scale human efforts rather than attempting to replace human judgment and local knowledge entirely.

As AI continues to evolve and be applied to social challenges, ongoing research, evaluation, and collaboration between technologists, domain experts, and affected communities will be essential to ensure that these technologies truly empower and benefit those they aim to serve [323]. The future of AI for social good holds great promise, but realizing this potential will require thoughtful design, implementation, and governance that prioritizes human values, ethical considerations, and sustainable impact.

One key aspect of empowering communities through AI for social good is the importance of participatory approaches and co-design. The case of the Coproduction Lab in Indonesia, a collaboration between UN Global Pulse and Pulse Lab Jakarta, illustrates this principle [264]. The lab works with local communities to co-create AI solutions for development challenges. In one project, they developed an AI-powered system to analyze social media data for early warning of food price fluctuations, which can significantly impact vulnerable populations. Involving local stakeholders in the design and implementation process ensured that the AI system addressed real community needs and was culturally appropriate. Community members reported feeling empowered to use technology to address local challenges and have a voice in the development of AI solutions that affect their lives.

This case highlights the importance of community involvement and local knowledge in developing AI solutions for social good. Adopting a participatory approach allows AI projects to empower communities to shape the technologies that impact them, leading to more effective and sustainable solutions [171].

The potential of AI to empower marginalized groups is demonstrated by the case of Safetipin, an AI-powered app designed to improve women's safety in urban areas [282]. The app uses machine learning algorithms to analyze user-generated data and environmental factors to create safety scores for different areas of a city. In Delhi, India, the implementation of Safetipin led to the improvement of over 7,800 dark spots identified by the app, enhancing the safety of public spaces. Women using the app reported feeling more empowered to navigate their cities safely and contribute to community safety efforts.

This case illustrates how AI can be leveraged to address gender-based vulnerabilities and empower women to take an active role in improving urban safety. Providing data-driven insights and facilitating community participation allows Safetipin to demonstrate the potential of AI to contribute to more inclusive and safe urban environments [359].

In the field of human rights, the case of the Syrian Archive project shows how AI can empower human rights defenders and journalists [19]. This initiative uses machine learning algorithms to verify and catalog video evidence of human rights violations in Syria. The AI system has helped process over 3 million videos, significantly accelerating the documentation of human rights abuses. Human rights researchers reported feeling more empowered to build comprehensive cases and advocate for justice, with the AI system handling the time-consuming task of video verification and categorization.

This case demonstrates how AI can augment human capabilities in human rights documentation and advocacy. Automating the processing of large volumes of video evidence enables AI to allow human rights defenders to focus on analysis, reporting, and strategic advocacy efforts [20].

The application of AI in improving access to legal services for underserved populations offers another example of empowerment for social good. The case of JusticeBot, an AI-powered chatbot developed by Stanford Law School's Legal Design Lab, illustrates this potential [182]. JusticeBot uses natural language processing to provide basic legal information and guidance to individuals who cannot afford traditional legal services. In a pilot study in California, users of JusticeBot reported a 40% increase in their understanding of legal processes and felt more empowered to navigate the legal system.

This case highlights how AI can democratize access to legal information and empower individuals to understand and assert their legal rights. Providing easily accessible legal guidance allows AI tools like JusticeBot to help bridge the justice gap for underserved communities [211].

In conclusion, these case studies demonstrate the significant potential of AI to empower individuals and communities in addressing complex social challenges. From environmental conservation and public health to social justice and human rights, AI technologies are showing promise in augmenting human efforts and scaling solutions for social good. However, it is crucial to approach the development and deployment of AI for social impact with careful consideration of ethical implications, community participation, and long-term sustainability.

The most successful implementations of AI for social good maintain a human-centered approach, leveraging AI capabilities to enhance and scale human efforts rather than replacing human judgment and local knowledge. They also prioritize inclusivity, ensuring that the benefits of AI are accessible to marginalized and underserved populations.

As we continue to explore and expand the applications of AI for social good, it will be essential to foster collaboration between technologists, domain experts, policymakers, and affected communities. This interdisciplinary approach can help ensure that AI solutions are ethically developed, culturally appropriate, and truly empowering for the communities they aim to serve [342].

Furthermore, as these AI systems become more prevalent in addressing social challenges, it will be crucial to develop frameworks for measuring and evaluating their long-term impact. This includes not only assessing the immediate outcomes of AI interventions but also considering their broader societal implications and potential unintended consequences [105].

The future of AI for social good holds great promise for empowering individuals and communities to address pressing global challenges. Continuing to innovate, learn from successes and failures, and prioritize ethical and inclusive approaches allows us to harness the potential of AI to create more equitable, sustainable, and empowered societies. As we move forward, it will be essential to maintain a balance between technological innovation and human values, ensuring that AI remains a tool for empowerment and positive social change [267].

V

Frontiers of Human-AI Empowerment

As we look toward the future of Human-AI Empowerment, we find ourselves at the cusp of transformative technological advancements and their profound implications for human capabilities and societal structures. This section explores emerging technologies and their potential for empowerment, addresses the challenges in scaling Human-AI Empowerment initiatives, examines the role of policy and governance in shaping AI empowerment, discusses future research directions, and envisions a future of empowered human-AI collaboration. Critically examining these frontiers aims to provide a roadmap for researchers, policymakers, and practitioners to navigate the complex landscape of Human-AI Empowerment in the coming years.

5.1 EMERGING TECHNOLOGIES AND THEIR POTENTIAL FOR EMPOWERMENT

The rapid pace of technological innovation continues to unveil new possibilities for Human-AI Empowerment. This subsection examines several emerging technologies that hold significant potential for enhancing human capabilities and expanding opportunities for empowerment.

One of the most promising frontiers in Human-AI Empowerment is the development of brain-computer interfaces (BCIs). BCIs offer the potential to create direct communication pathways between the human brain and external devices, opening up new avenues for cognitive enhancement and assistive technologies [354]. For instance, researchers at the University of California, San Francisco, have demonstrated a BCI system capable of decoding speech directly from brain signals, potentially enabling communication for individuals with severe speech impairments [18]. As BCI technology advances, it could dramatically empower individuals with disabilities, enhance cognitive capabilities, and create novel forms of human-AI interaction.

The continued evolution of augmented reality (AR) and virtual reality (VR) technologies presents another frontier for Human-AI Empowerment. These immersive technologies, when combined with AI, have the potential to revolutionize fields such as education, healthcare, and professional training [37]. For example, Microsoft's HoloLens 2, an AR headset, is being used in surgical settings to provide real-time, AI-enhanced visualization of patient anatomy during procedures [219]. As AR and VR technologies become more sophisticated and integrated with AI systems, they could offer unprecedented opportunities for experiential learning, remote collaboration, and enhanced perception of the world around us.

The emergence of quantum computing represents a potential paradigm shift in AI capabilities and, by extension, Human-AI Empowerment. Quantum computers have the potential to solve complex problems that are currently intractable for classical computers, potentially leading to breakthroughs in areas such as drug discovery, financial modeling, and climate prediction [261]. For instance, researchers at Google have demonstrated quantum supremacy, performing a calculation in 200 seconds that would take the world's most powerful supercomputer 10,000 years [22]. As quantum computing matures and becomes more accessible, it could dramatically enhance our ability to tackle complex global challenges and empower individuals with unprecedented computational resources.

The field of neuroengineering, which combines neuroscience with engineering principles, is another frontier with significant implications for Human-AI Empowerment. Advances in this field could lead to the development of neural prosthetics that restore or enhance sensory, motor, and cognitive functions [189]. For example, researchers at Johns Hopkins University have developed a prosthetic arm controlled by neural signals that provides sensory feedback, allowing users to feel texture and pressure [104]. As neuroengineering technologies progress, they could offer transformative empowerment for individuals with neurological disorders and potentially enhance cognitive and physical capabilities for the broader population.

The development of advanced natural language processing (NLP) and generation technologies, exemplified by models like GPT-3, represents another frontier in Human-AI Empowerment [48]. These large language models demonstrate remarkable capabilities in understanding and generating human-like text, potentially revolutionizing how we interact with information and knowledge. For instance, GPT-3 has been used to assist in creative writing, code generation, and even scientific research [314]. As these technologies continue to evolve, they could dramatically enhance human cognitive capabilities, enabling more efficient information processing, idea generation, and problem-solving across various domains.

The field of synthetic biology, which combines biology with engineering principles, presents exciting possibilities for Human-AI Empowerment. AI-driven tools are increasingly being used to design and engineer biological systems, potentially leading to breakthroughs in areas such as personalized medicine, biofuel production, and environmental remediation [56]. For example, researchers at MIT have used machine learning to design new antibiotics that can combat drug-resistant bacteria [307]. As synthetic biology and AI continue to converge, they could empower individuals with unprecedented control over biological processes and open up new frontiers in healthcare and biotechnology.

The development of advanced robotics, particularly in the realm of human-robot collaboration, represents another important frontier for Human-AI Empowerment. Collaborative robots, or "cobots," are designed to work alongside humans, augmenting our physical capabilities and potentially transforming various industries [10]. For instance, researchers at ETH Zurich have developed a robotic exoskeleton that can be controlled by thought, potentially empowering individuals with mobility impairments [225]. As robotic technologies become more sophisticated and integrated with AI systems, they could dramatically enhance human physical capabilities and create new forms of human-machine synergy.

The emergence of edge AI, which involves deploying AI algorithms on local devices rather than in the cloud, presents new opportunities for Human-AI Empowerment. Edge AI has the potential to enhance privacy, reduce latency, and enable AI-powered applications in resource-constrained environments [367]. For example, Google's Live Transcribe app uses on-device machine learning to provide real-time speech-to-text transcription for deaf and hard-of-hearing individuals, even without an internet connection [126]. As edge AI technologies advance, they could empower individuals with more personalized, responsive, and privacy-preserving AI assistance in their daily lives.

These emerging technologies, while promising, also raise important ethical and societal questions that must be carefully considered as we move forward. Issues of privacy, security, equity, and the potential for unintended consequences must be addressed to ensure that these technologies truly empower all individuals and contribute to societal well-being [106].

As we stand at the frontier of these technological advancements, it is crucial to approach their development and deployment with a human-centered perspective, ensuring that they augment and empower human capabilities rather than replace or diminish them. The challenge lies in harnessing these technologies to create a future where AI and humans work in symbiosis, each enhancing the capabilities of the other to address complex global challenges and unlock new realms of human potential [298].

5.2 CHALLENGES IN SCALING HUMAN-AI EMPOWERMENT

While the potential of Human-AI Empowerment is vast, scaling these initiatives to benefit larger populations and diverse contexts presents significant challenges. This subsection examines the key obstacles that must be addressed to realize the full potential of Human-AI Empowerment on a global scale. These interrelated challenges, visualized in Figure 5.1, span technological, social, ethical, and economic dimensions.

One of the primary challenges in scaling Human-AI Empowerment is the issue of accessibility and the digital divide. Despite rapid technological advancements, a significant portion of the global population still lacks access to basic digital infrastructure and technologies [335]. This disparity in access to AI technologies and the skills needed to leverage them effectively could exacerbate existing social and economic inequalities. For instance, a study by the International Telecommunication Union found that only 19% of individuals in least developed countries use the internet, compared to 87% in developed countries [332]. Addressing this digital divide is crucial to ensure that the benefits of Human-AI Empowerment are equitably distributed and do not become a source of further societal stratification.

The challenge of data quality and representativeness presents another significant hurdle in scaling Human-AI Empowerment. AI systems are only as good as the data they

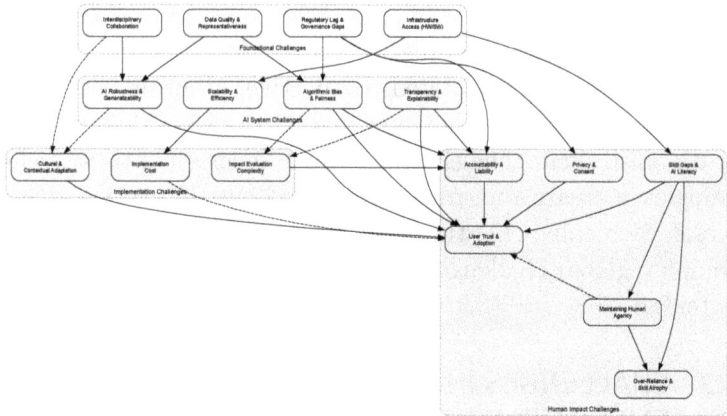

Figure 5.1 Interrelated Challenges in Scaling Human-AI Empowerment Initiatives

are trained on, and biases in training data can lead to biased or discriminatory outcomes when these systems are deployed at scale [214]. For example, a study by Buolamwini and Gebru found that commercial gender classification systems had much higher error rates for darker-skinned females compared to lighter-skinned males [51]. Ensuring diverse and representative datasets, particularly for underrepresented populations, is essential for creating AI systems that empower all individuals equitably.

The scalability of AI systems themselves poses a significant challenge. As AI applications grow in complexity and scale, they often require increasingly large amounts of computational resources and energy. This raises concerns about the environmental sustainability of large-scale AI deployment [309]. For instance, training a single large language model can emit as much carbon as five cars in their lifetimes [309]. Developing more efficient AI algorithms and hardware, as well as leveraging renewable energy sources for AI computation, will be crucial for sustainable scaling of Human-AI Empowerment initiatives.

Another challenge lies in the adaptability of AI systems to diverse cultural and linguistic contexts. Many AI systems are developed primarily in English and Western cultural contexts, limiting their effectiveness and relevance in other parts of the world [284]. For example, a study by Nekoto et al. highlighted the severe underrepresentation of African languages in natural language processing research and applications [229]. Scaling Human-AI Empowerment globally will require significant efforts to develop AI systems that are culturally and linguistically adaptable and relevant to diverse populations.

The ethical implications of AI systems become increasingly complex as they are scaled up and integrated more deeply into societal structures. Issues of privacy, consent, and autonomy become more pressing as AI systems gain access to larger amounts of personal data and play more significant roles in decision-making processes [222]. For instance, the use of facial recognition technologies by law enforcement agencies has raised significant concerns about privacy and civil liberties [118]. Developing robust ethical frameworks and governance structures that can keep pace with rapidly scaling AI technologies is a crucial challenge.

The need for interdisciplinary collaboration presents another challenge in scaling Human-AI Empowerment. Effective development and deployment of AI systems for empowerment often require expertise from various fields, including computer science, psychology, ethics, and domain-specific knowledge [267]. However, bridging these disciplinary divides and fostering effective collaboration can be challenging. For example, a survey of AI researchers found that only 34% regularly collaborate with scholars from other disciplines [127]. Overcoming these silos and fostering truly interdisciplinary approaches will be crucial for addressing the complex challenges of scaling Human-AI Empowerment.

The challenge of user trust and adoption is particularly salient as AI systems are deployed at scale. Despite the potential benefits, many individuals may be hesitant to trust or adopt AI technologies, particularly in sensitive domains such as healthcare or financial decision-making [325]. A survey by the Pew Research Center found that 58% of Americans feel that computer programs will always reflect some level of human bias [303]. Building trust through transparency, explainability, and demonstrated reliability of AI systems will be crucial for their widespread adoption and effective empowerment of users.

The regulatory and legal challenges associated with scaling Human-AI Empowerment are significant. As AI systems become more pervasive and influential, existing legal frameworks may struggle to address new issues of liability, intellectual property, and rights in human-AI interactions [55]. For instance, the question of who is liable when an AI system makes a decision that results in harm remains a complex legal issue [174]. Developing appropriate regulatory frameworks that can balance innovation with protection of individual rights and societal interests is a key challenge in scaling Human-AI Empowerment.

Lastly, the challenge of maintaining human agency and avoiding over-reliance on AI systems is crucial as these technologies scale. While AI can significantly enhance human capabilities, there is a risk of individuals becoming overly dependent on AI assistance, potentially leading to atrophy of certain cognitive or physical skills [57]. For example, over-reliance on GPS navigation has been shown to negatively impact spatial memory and navigation skills [228]. Striking the right balance between AI assistance and human skill development will be essential for true empowerment as these technologies scale.

Addressing these challenges will require concerted efforts from researchers, policymakers, industry leaders, and civil society. It will involve not only technological innovations but also social, economic, and policy interventions to ensure that the scaling of Human-AI Empowerment initiatives truly benefits all of humanity. As we navigate these challenges, maintaining a focus on human values, ethical considerations, and the ultimate goal of enhancing human capabilities and well-being will be paramount [106].

5.3 THE ROLE OF POLICY AND GOVERNANCE IN SHAPING AI EMPOWERMENT

The development and deployment of AI technologies for human empowerment do not occur in a vacuum but are deeply influenced by policy and governance frameworks. This subsection examines the crucial role that policy and governance play in shaping the trajectory of Human-AI Empowerment and ensuring that it aligns with societal values and ethical principles.

One of the primary roles of policy in AI empowerment is to promote innovation while safeguarding public interests. Governments around the world are grappling with how to foster AI development and adoption while addressing concerns about privacy, security, and fairness [59]. For instance, the European Union's General Data Protection Regulation (GDPR) has set a global standard for data protection and privacy in the digital age, significantly impacting how AI systems can collect and use personal data [345]. Such regulations play a crucial role in ensuring that AI empowerment initiatives respect individual rights and operate within ethical boundaries.

Policy also plays a critical role in addressing the potential socioeconomic impacts of AI, particularly in the realm of employment and workforce transformation. As AI technologies continue to automate various tasks and reshape industries, policies are needed to support workforce adaptation and ensure that the benefits of AI are broadly shared [4]. For example, Singapore's SkillsFuture initiative provides citizens with resources for lifelong learning and reskilling, helping to prepare the workforce for an AI-driven economy [301]. Such policies are essential for ensuring that AI empowerment extends to all sectors of society and does not exacerbate existing inequalities.

The governance of AI research and development is another crucial area where policy plays a significant role. Ensuring responsible AI development requires frameworks for ethical review, transparency, and accountability [160]. For instance,

Canada's Directive on Automated Decision-Making sets out requirements for impact assessments, transparency, and human oversight in the use of AI systems in government services [240]. Such governance frameworks are essential for building public trust in AI technologies and ensuring their alignment with societal values.

International cooperation and policy harmonization are increasingly important as AI technologies transcend national borders. Initiatives like the OECD AI Principles, adopted by 42 countries, aim to promote AI that is innovative, trustworthy, and respects human rights and democratic values [239]. Such international efforts are crucial for creating a global ecosystem that fosters responsible AI development and empowerment while addressing challenges like data flows, algorithmic transparency, and cross-border AI services.

Policy also plays a vital role in promoting AI literacy and public engagement. As AI becomes more pervasive, it is crucial that the general public has a basic understanding of AI technologies and their implications [198]. For example, Finland's "Elements of AI" course, offered free to all EU citizens, aims to demystify AI and empower individuals to participate in discussions about its societal impact [241]. Such initiatives are essential for fostering informed public discourse and ensuring that AI empowerment is shaped by diverse perspectives.

The governance of AI in critical sectors such as healthcare, finance, and criminal justice requires particular attention due to the high stakes involved. Policies are needed to ensure that AI systems in these domains are safe, reliable, and fair [222]. For instance, the U.S. Food and Drug Administration has developed a regulatory framework for AI-based medical devices, addressing issues of safety, efficacy, and ongoing performance monitoring [108]. Such sector-specific governance is crucial for realizing the potential of AI empowerment while mitigating risks in sensitive domains.

The role of policy in addressing AI bias and promoting fairness is increasingly recognized as crucial for ethical AI empowerment. Policies and guidelines are being developed to ensure that AI systems do not perpetuate or exacerbate existing societal biases [214]. For example, the AI Fairness 360 toolkit, developed by IBM, provides resources for detecting and mitigating bias in AI systems [150]. Such initiatives, supported by appropriate policies, are essential for ensuring that AI empowerment benefits all individuals equitably.

Governance frameworks for AI transparency and explainability are another critical area where policy plays a key role. As AI systems become more complex and influential, there is a growing need for mechanisms to understand and audit their decision-making processes [85]. The EU's proposed AI Act, for instance, includes requirements for high-risk AI systems to be sufficiently transparent and provide appropriate levels of human oversight [68]. Such policies are crucial for maintaining accountability and trust in AI systems as they become more integrated into critical decision-making processes.

The governance of AI in the context of global challenges, such as climate change and public health crises, is an emerging area where policy plays a vital role. Frameworks are needed to guide the development and deployment of AI solutions for these complex, transnational issues while ensuring ethical considerations and equitable benefits [342]. For example, the UN's AI for Good initiative aims to harness AI technologies to accelerate progress toward the Sustainable Development Goals [333]. Such policy frameworks are essential for directing AI development toward addressing pressing global challenges and empowering humanity as a whole.

Privacy-preserving AI technologies, such as federated learning and differential privacy, present new challenges and opportunities for policy and governance [164]. While these technologies offer promising solutions for data privacy concerns, they also require new regulatory approaches to ensure their effective implementation and oversight. For instance, the U.S. National Institute of Standards and Technology (NIST) is developing guidelines for privacy-enhancing technologies in AI systems [242]. Such policy efforts are crucial for balancing the benefits of AI empowerment with the protection of individual privacy rights.

The governance of AI in warfare and national security contexts raises particularly complex ethical and policy challenges [146]. While AI has the potential to enhance defensive capabilities and reduce human casualties, it also raises concerns about autonomous weapons systems and the potential for AI-enabled conflicts. International efforts, such as the UN's Group of Governmental Experts on Lethal Autonomous Weapons Systems, are working to develop governance frameworks for military AI applications [23]. These policy initiatives are critical for ensuring that AI empowerment in military contexts aligns with international humanitarian law and ethical principles.

Lastly, the role of policy in fostering public-private partnerships and collaborative governance models for AI is increasingly recognized as crucial [94]. As AI development often occurs in the private sector, while its impacts are felt across society, frameworks for collaboration between industry, government, academia, and civil society are essential. For example, the Partnership on AI brings together diverse stakeholders to develop best practices for AI technologies [243]. Such collaborative governance models, supported by appropriate policies, are vital for ensuring that AI empowerment initiatives benefit from diverse perspectives and align with broader societal interests.

In conclusion, policy and governance play a multifaceted and crucial role in shaping the trajectory of Human-AI Empowerment. From promoting innovation and safeguarding public interests to addressing socioeconomic impacts and ensuring ethical development, policy frameworks are essential for realizing the potential of AI while mitigating its risks. As AI technologies continue to evolve and permeate various aspects of society, the development of adaptive, inclusive, and forward-looking governance structures will be paramount [59].

The challenge lies in creating policies that are flexible enough to accommodate rapid technological advancements while being robust enough to protect fundamental human rights and societal values. This will require ongoing dialogue between policymakers, technologists, ethicists, and the public to navigate the complex landscape of AI governance [106]. As we move forward, it will be crucial to develop governance frameworks that not only regulate AI but actively steer its development toward empowering humanity and addressing global challenges. In doing so, we can work toward a future where AI truly serves as a tool for human flourishing and societal progress [298].

5.4 FUTURE RESEARCH DIRECTIONS IN HUMAN-AI EMPOWERMENT

As the field of Human-AI Empowerment continues to evolve, several key areas emerge as critical for future research. This subsection explores potential research directions that could significantly advance our understanding and implementation of AI systems for human empowerment.

One crucial area for future research is the development of more sophisticated models of human-AI collaboration. While current research often focuses on either AI automation or human-in-the-loop systems, there is a need for more nuanced models that capture the dynamic interplay between human and artificial intelligence [267]. Future research could explore adaptive collaboration models where the division of tasks between humans and AI evolves based on context, user capabilities, and system performance. For instance, studies could investigate how to design AI systems that seamlessly transition between supportive and leading roles in problem-solving tasks, optimizing for both task performance and human skill development [165].

Another important research direction is the exploration of long-term cognitive and social impacts of Human-AI Empowerment. As AI systems become more integrated into our daily lives and decision-making processes, it is crucial to understand their effects on human cognition, behavior, and social dynamics over extended periods [57]. Longitudinal studies examining how prolonged interaction with AI systems affects cognitive skills, decision-making patterns, and social relationships could provide valuable insights for designing empowering AI technologies. For example, research could investigate how AI-assisted learning tools impact long-term knowledge retention and problem-solving abilities [136].

The development of more robust and generalizable AI systems for empowerment presents another critical research challenge. Current AI systems often excel in narrow, well-defined domains but struggle with tasks requiring common sense reasoning or adaptability to novel situations [208]. Future research could focus on developing AI architectures that can better generalize across different contexts and tasks, more closely mimicking human cognitive flexibility. This could include exploring hybrid AI systems that combine deep learning with symbolic reasoning, or investigating meta-learning approaches that enable AI systems to learn how to learn more efficiently [183].

Research into enhancing the interpretability and explainability of AI systems is crucial for fostering trust and effective human-AI collaboration. As AI systems become more complex, ensuring that their decision-making processes are transparent and understandable to human users becomes increasingly challenging [85]. Future research could explore novel visualization techniques, natural language explanation methods, and interactive explanatory interfaces that make AI reasoning more accessible to users with varying levels of technical expertise. For instance, studies could investigate how to generate personalized, context-aware explanations that adapt to the user's knowledge level and information needs [220].

The ethical implications of Human-AI Empowerment, particularly in terms of autonomy, privacy, and fairness, require ongoing research attention. As AI systems take on more significant roles in decision-making and personal assistance, questions arise about their impact on human agency and the potential for manipulation or undue influence [106]. Future research could explore frameworks for ethical AI design that prioritize user empowerment and autonomy, investigate methods for preserving privacy in highly personalized AI systems, and develop more sophisticated approaches to detecting and mitigating bias in AI-assisted decision-making [214].

Research into the psychological and cognitive factors that influence human trust and acceptance of AI systems is another important direction. Understanding how humans perceive, trust, and interact with AI technologies is crucial for designing systems that can effectively empower users [325]. Future studies could investigate the factors that contribute to appropriate trust in AI systems, explore methods for calibrating user confidence in AI recommendations, and examine how cultural and individual differences affect human-AI interaction patterns. This research could inform the development of AI systems that are more intuitive, trustworthy, and aligned with human values and expectations [75].

The intersection of Human-AI Empowerment with emerging technologies such as brain-computer interfaces, augmented reality, and quantum computing presents exciting opportunities for future research. Investigating how these technologies can be synergistically combined with AI to enhance human capabilities could lead to breakthrough empowerment applications [354]. For example, research could explore how brain-computer interfaces could be used to create more intuitive and responsive AI assistants, or how quantum machine learning could enable new forms of cognitive enhancement [36].

Research into scalable and sustainable approaches to Human-AI Empowerment is crucial as we look toward global implementation. This includes investigating energy-efficient AI architectures, developing AI systems that can operate effectively in resource-constrained environments, and exploring methods for adapting AI technologies to diverse cultural and linguistic contexts [309]. Future studies could also examine sustainable models for AI education and literacy, ensuring that the benefits of AI empowerment are accessible to diverse populations globally [198].

Finally, interdisciplinary research integrating insights from cognitive science, neuroscience, and AI could lead to more effective and human-centered empowerment technologies. Deepening our understanding of human cognition and learning processes allows us to design AI systems that more naturally augment and enhance human intelligence [137]. Future research could explore how findings from cognitive neuroscience can inform the development of AI systems that better align with human cognitive architectures, or investigate how AI can be used to enhance specific cognitive functions based on neuroscientific insights.

In conclusion, the future research directions in Human-AI Empowerment are diverse and multifaceted, reflecting the complexity and potential of this field. As we move forward, it will be crucial to adopt interdisciplinary approaches, combining insights from computer science, cognitive science, psychology, ethics, and other relevant disciplines. Pursuing these research directions enables us to work toward developing AI technologies that truly empower humans, enhancing our capabilities while respecting our values and autonomy. The ultimate goal is to create a symbiotic relationship between humans and AI, where each augments the other's strengths, leading to unprecedented levels of problem-solving ability and creative potential [298].

5.5 ENVISIONING A FUTURE OF EMPOWERED HUMAN-AI COLLABORATION

As we stand at the frontier of Human-AI Empowerment, it is crucial to envision a future where humans and AI systems collaborate synergistically, enhancing each other's capabilities and working together to address complex global challenges. This subsection explores a vision of this empowered future, considering potential scenarios, opportunities, and considerations for realizing this vision.

In this envisioned future, AI systems have evolved to become true cognitive partners, seamlessly integrating with human thought processes and activities across various domains [58]. Imagine a world where personalized AI assistants, deeply attuned to individual cognitive styles and preferences, collaborate with humans on complex problem-solving tasks. These AI partners could offer real-time insights, suggest novel approaches, and even anticipate needs before they are explicitly expressed. For instance, in scientific research, AI systems might work alongside human scientists, generating hypotheses, designing experiments, and interpreting results, dramatically accelerating the pace of discovery [175].

Education in this future scenario is transformed by AI-empowered learning environments that adapt in real-time to each student's needs, learning style, and pace [143]. Imagine virtual tutors that can explain concepts in multiple ways, provide personalized examples, and create engaging, interactive learning experiences. These AI-enhanced educational tools could help bridge achievement gaps, make high-quality education accessible to all, and foster a culture of lifelong learning. Furthermore, AI systems could help individuals identify and develop their unique talents and passions, guiding them toward fulfilling career paths that align with their abilities and interests [123].

In healthcare, the collaboration between AI systems and human medical professionals could lead to unprecedented levels of personalized and preventive care [324]. Envision AI systems that continuously monitor an individual's health data, predict potential issues before they manifest, and work with healthcare providers to develop tailored treatment plans. These systems could also assist in complex diagnoses, suggest innovative treatment approaches, and even perform intricate surgical procedures in collaboration with human surgeons, potentially saving countless lives and improving global health outcomes.

The workplace of the future could see humans and AI systems working side by side in harmony, each leveraging their unique strengths [73]. AI systems could handle routine tasks, data analysis, and complex computations, freeing humans to focus on creative problem-solving, strategic thinking, and interpersonal interactions. This collaboration could lead to unprecedented levels of productivity and innovation across industries. Moreover, AI systems could help create more inclusive and equitable work environments by mitigating human biases in hiring, promotion, and decision-making processes.

In the realm of creativity and arts, AI could become a powerful tool for augmenting human imagination and expanding the boundaries of artistic expression [92]. Imagine AI systems that can generate novel musical compositions, suggest unique color palettes for paintings, or even co-author stories with human writers. These AI collaborators could inspire artists to explore new creative directions and push the limits of their craft, potentially giving rise to entirely new art forms and modes of expression.

Environmental stewardship and sustainable development could be significantly enhanced through empowered human-AI collaboration [342]. AI systems could work with human experts to model complex ecological systems, predict the impacts of various interventions, and optimize resource allocation for maximum sustainability. This collaboration could lead to innovative solutions for climate change mitigation, biodiversity conservation, and sustainable urban planning, helping to create a more environmentally resilient future. This synergistic relationship can create a virtuous cycle of empowerment, as depicted in Figure 5.2, where enhanced human capabilities lead to better AI systems, which in turn further empower humans.

Governance and policy-making could be transformed by AI systems that can process vast amounts of data, model complex societal dynamics, and predict the long-term impacts of various policy options [340]. Human policymakers, empowered by these AI insights, could make more informed, equitable, and effective decisions. Moreover, AI systems could enhance civic participation by providing citizens with personalized information about policy issues and facilitating more direct forms of democratic engagement.

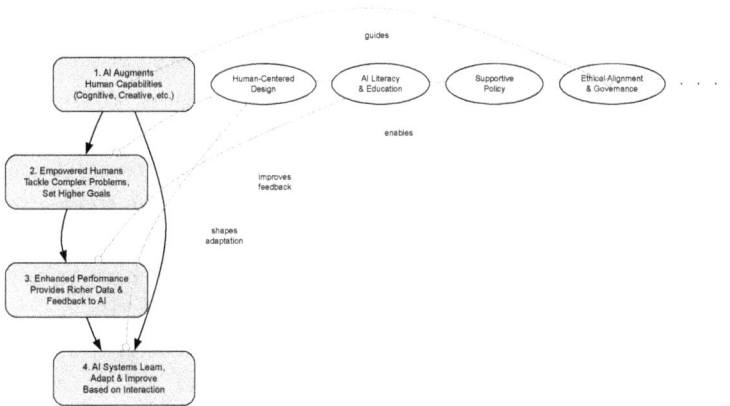

Figure 5.2 The Virtuous Cycle of Empowered Human-AI Collaboration

However, realizing this vision of empowered human-AI collaboration requires careful consideration of potential challenges and ethical implications. Issues of privacy, autonomy, and the digital divide must be addressed to ensure that the benefits of AI empowerment are equitably distributed and do not come at the cost of fundamental human rights [106]. It will be crucial to develop robust governance frameworks that can adapt to rapidly evolving AI technologies while safeguarding human values and societal well-being.

Furthermore, as AI systems become more integrated into our lives and decision-making processes, it will be essential to maintain a balance between AI assistance and human agency [298]. The goal should be to use AI to enhance human capabilities and decision-making, not to replace human judgment entirely. This will require ongoing efforts in AI education and literacy, ensuring that individuals can critically evaluate AI recommendations and maintain meaningful control over important life decisions.

In conclusion, the future of empowered human-AI collaboration holds immense potential for addressing global challenges, enhancing human capabilities, and creating a more sustainable and equitable world. Leveraging the complementary strengths of human and artificial intelligence allows us to potentially achieve levels of problem-solving, creativity, and innovation that were previously unimaginable. However, realizing this vision will require concerted efforts from researchers, policymakers, industry leaders, and society at large to ensure that AI technologies are developed and deployed in ways that truly empower humanity and align with our collective values and aspirations [267].

As we move toward this future, it will be crucial to maintain a human-centered approach, viewing AI not as a replacement for human intelligence, but as a powerful tool for augmenting and amplifying human potential. This approach allows us to work toward a future where the synergy between human and artificial intelligence leads to unprecedented advancements in science, technology, arts, and social progress, ultimately contributing to the flourishing of humanity as a whole [320].

CHAPTER 6

Conclusion

As we conclude this comprehensive exploration of Human-AI Empowerment, it is evident that we stand at a pivotal moment in the co-evolution of human and artificial intelligence. The rapid advancements in AI technologies, coupled with our deepening understanding of human cognition and societal dynamics, present unprecedented opportunities for enhancing human capabilities and addressing complex global challenges. However, this potential for empowerment is accompanied by significant ethical, social, and technical challenges that require careful consideration and proactive management.

Throughout this book, we have examined the multifaceted nature of Human-AI Empowerment, from its theoretical foundations to practical applications across various domains. We have explored emerging technologies, methodological approaches, and case studies that illustrate the transformative potential of human-AI collaboration. As we reflect on these insights, several key themes emerge that will shape the future trajectory of Human-AI Empowerment:

6.1 THE SYMBIOTIC RELATIONSHIP BETWEEN HUMAN AND ARTIFICIAL INTELLIGENCE

The most promising path forward for Human-AI Empowerment lies not in the replacement of human intelligence by AI,

DOI: 10.1201/9781003536628-6

but in the cultivation of a symbiotic relationship between the two. As demonstrated in domains such as healthcare, education, and scientific research, the combination of human creativity, contextual understanding, and ethical judgment with AI's computational power and pattern recognition capabilities can lead to outcomes that surpass what either could achieve alone [267].

For instance, in healthcare, we have seen how AI-powered diagnostic tools can augment the capabilities of human physicians, leading to more accurate and timely diagnoses. The case of IBM's Watson for Oncology, which analyzes vast amounts of medical literature and patient data to suggest treatment options, exemplifies this symbiosis [304]. However, the system's recommendations are most effective when combined with the clinical judgment and empathetic care of human doctors.

In the realm of scientific research, AI systems like AlphaFold have demonstrated remarkable capabilities in predicting protein structures, a critical task in understanding diseases and developing new treatments [162]. Yet, the true power of these AI breakthroughs is realized when human scientists leverage these insights to drive forward their research, asking new questions and exploring novel hypotheses that the AI alone could not generate.

Future developments in Human-AI Empowerment should focus on enhancing this complementarity, designing AI systems that augment human strengths while compensating for cognitive limitations. This could involve developing more sophisticated human-AI interfaces that allow for intuitive collaboration, or creating AI systems that can adapt their level of assistance based on the user's expertise and the complexity of the task at hand.

6.2 THE IMPERATIVE OF ETHICAL AI DEVELOPMENT

As AI systems become more sophisticated and pervasive, the ethical implications of their development and deployment become increasingly critical. Issues of privacy, fairness, transparency, and accountability must be at the forefront of Human-AI Empowerment initiatives [106]. The case studies and research discussed in this book highlight the potential for AI to exacerbate existing societal biases or create new forms of inequality if not carefully designed and implemented.

For example, the use of AI in criminal justice systems has raised significant concerns about bias and fairness. The COMPAS algorithm, used in some U.S. states to assess recidivism risk, has been criticized for potentially perpetuating racial biases in the criminal justice system [17]. This case underscores the critical need for rigorous testing, ongoing monitoring, and transparent reporting of AI systems, especially in high-stakes domains.

In the field of facial recognition, concerns about privacy and consent have come to the forefront. The case of Clearview AI, which scraped billions of images from social media platforms to create a facial recognition database, has sparked intense debate about the ethical boundaries of AI development and deployment [142].

Moving forward, it is essential to integrate ethical considerations into every stage of AI development, from initial conceptualization to real-world deployment and ongoing monitoring. This could involve developing more robust frameworks for ethical AI design, such as the IEEE's Ethically Aligned Design guidelines [244], and implementing these principles in practical development processes.

Furthermore, the development of AI ethics review boards, similar to institutional review boards in medical research, could provide a mechanism for independent oversight and ethical evaluation of AI projects. These boards could assess potential risks, ensure compliance with ethical guidelines, and promote responsible innovation in Human-AI Empowerment initiatives.

6.3 THE NEED FOR INTERDISCIPLINARY COLLABORATION

The complex nature of Human-AI Empowerment necessitates collaboration across diverse disciplines. As we have seen, insights from computer science, cognitive psychology, neuroscience, ethics, and social sciences all play crucial roles in developing effective and responsible AI systems for human empowerment [137].

For instance, the development of brain-computer interfaces (BCIs) for empowering individuals with severe motor disabilities requires expertise from neuroscientists to understand brain function, engineers to design the hardware, computer scientists to develop the algorithms, ethicists to address privacy and agency concerns, and clinicians to implement and evaluate the technology in real-world settings [354].

Similarly, the creation of AI-powered educational tools that truly enhance learning outcomes requires input from cognitive scientists to understand how people learn, educators to identify pedagogical needs, data scientists to develop adaptive algorithms, and ethicists to ensure the protection of student privacy and well-being [143].

Future research and development in this field should prioritize interdisciplinary approaches, fostering dialogue and collaboration between technologists, domain experts, policymakers, and ethicists to ensure a holistic and balanced approach to Human-AI Empowerment. This could involve creating dedicated interdisciplinary research centers focused on Human-AI Empowerment, developing cross-disciplinary curricula in higher education, and establishing funding mechanisms that explicitly require interdisciplinary collaboration.

6.4 THE IMPORTANCE OF HUMAN-CENTERED DESIGN

Throughout our exploration of Human-AI Empowerment, the importance of human-centered design has been a recurring theme. AI systems designed for empowerment must be intuitive, transparent, and aligned with human cognitive processes and social dynamics [298]. This requires a deep understanding of human needs, capabilities, and limitations, as well as ongoing user engagement and feedback in the development process.

The success of virtual assistants like Apple's Siri or Amazon's Alexa in empowering users to perform various tasks through natural language interaction demonstrates the importance of human-centered design. These systems have been iteratively refined based on user feedback and interaction data to become more intuitive and aligned with human communication patterns [147].

In contrast, the initial challenges faced by Google Glass, a wearable AI-powered device, highlight the consequences of neglecting human-centered design principles. The product faced significant backlash due to privacy concerns and social awkwardness, underscoring the need to consider not just technological capabilities, but also social norms and user comfort in designing AI-empowered devices [285].

Future advancements in Human-AI Empowerment should prioritize user experience and human factors alongside technological innovation. This could involve developing more sophisticated user research methodologies tailored to AI interactions, creating guidelines for explainable AI that ensure systems can effectively communicate their reasoning to users, and implementing participatory design processes that involve end-users throughout the development cycle.

6.5 THE CHALLENGE OF SCALABILITY AND ACCESSIBILITY

While many of the Human-AI Empowerment initiatives discussed in this book show great promise, a significant challenge lies in scaling these solutions to benefit larger populations and diverse contexts. Issues of digital divide, technological literacy, and resource constraints must be addressed to ensure that the benefits of AI empowerment are equitably distributed [335].

For example, while AI-powered personalized learning platforms have shown promising results in improving educational outcomes, their implementation often requires reliable internet access, compatible devices, and a certain level of digital literacy among students and teachers. These requirements can limit the accessibility of such tools in resource-constrained environments or among disadvantaged populations [270].

Similarly, the potential of AI in healthcare to empower patients and clinicians is significant, but realizing this potential globally requires addressing issues of data privacy, interoperability of health systems, and equitable access to technology [347].

Future efforts should focus on developing AI solutions that are not only powerful but also accessible, adaptable, and sustainable across various socioeconomic and cultural contexts. This could involve creating AI systems that can operate effectively with limited computational resources or intermittent internet connectivity, developing culturally adaptive AI interfaces, and implementing capacity-building programs to enhance AI literacy in underserved communities.

6.6   THE ROLE OF POLICY AND GOVERNANCE

As AI technologies continue to evolve and permeate various aspects of society, the role of policy and governance in shaping Human-AI Empowerment becomes increasingly critical. The development of adaptive, forward-looking regulatory frameworks that can keep pace with technological advancements while safeguarding human rights and societal values is a key challenge [59].

The European Union's General Data Protection Regulation (GDPR) and the proposed AI Act represent significant steps toward comprehensive governance of AI technologies [345][68]. These frameworks aim to protect individual rights, ensure transparency in AI decision-making, and promote responsible innovation. However, their implementation has also highlighted the challenges of balancing innovation with regulation and of enforcing such regulations in a rapidly evolving technological landscape.

In the United States, the development of AI governance has been more sector-specific, with agencies like the Food and Drug Administration (FDA) developing frameworks for regulating AI in healthcare [108]. This approach allows for more tailored regulation but may also lead to inconsistencies across sectors.

Future work in this area should focus on fostering international cooperation, developing flexible governance models that can adapt to technological advancements, and promoting public engagement in AI policy-making. This could involve creating international bodies for AI governance, similar to those that exist for climate change or nuclear non-proliferation, developing mechanisms for rapid policy experimentation and evaluation, and implementing public consultation processes to ensure AI governance reflects societal values and concerns.

6.7 THE POTENTIAL FOR ADDRESSING GLOBAL CHALLENGES

Human-AI Empowerment holds significant promise for addressing pressing global challenges, from climate change and healthcare to education and sustainable development. The case studies and research presented in this book demonstrate the potential for AI to enhance human problem-solving capabilities, facilitate more effective resource allocation, and enable novel approaches to complex issues [342].

For instance, in climate science, AI models are being used to improve climate predictions, optimize renewable energy systems, and develop more sustainable urban planning strategies [276]. These tools empower climate scientists and policymakers to make more informed decisions and develop more effective mitigation and adaptation strategies.

In global health, AI systems are being leveraged to predict disease outbreaks, accelerate drug discovery, and optimize resource allocation in healthcare systems [324]. These applications have the potential to significantly enhance our ability to respond to health crises and improve healthcare outcomes globally.

Future developments should explore how Human-AI Empowerment can be leveraged to accelerate progress toward the United Nations Sustainable Development Goals and other global priorities. This could involve developing AI systems specifically designed to address complex, interconnected global challenges, creating platforms for global collaboration on AI-empowered problem-solving, and implementing mechanisms to ensure that the benefits of these technologies are equitably distributed across the global population.

6.8 THE IMPERATIVE OF CONTINUOUS LEARNING AND ADAPTATION

The rapidly evolving nature of AI technologies necessitates a commitment to continuous learning and adaptation in the field of Human-AI Empowerment. As new capabilities emerge and unforeseen challenges arise, it will be crucial for researchers, practitioners, and policymakers to remain flexible and responsive.

This includes ongoing evaluation of the impacts of AI empowerment initiatives, regular reassessment of ethical guidelines, and the development of agile methodologies for AI development and deployment. For example, the field of autonomous vehicles has seen numerous iterations of technology, policy, and ethical frameworks as new challenges and capabilities have emerged [26].

Future efforts in this area could involve developing more sophisticated impact assessment methodologies for AI systems, creating mechanisms for rapid knowledge sharing across the AI community, and implementing adaptive regulatory frameworks that can evolve alongside technological advancements.

6.9   THE EXPLORATION OF LONG-TERM IMPACTS

While much of the current research focuses on the immediate or short-term effects of Human-AI Empowerment, it is crucial to consider the long-term implications of these technologies on human cognition, social structures, and cultural evolution. Future research should prioritize longitudinal studies that examine the cumulative effects of AI-enhanced capabilities on individual and societal development over extended periods [57].

For instance, as AI-powered personal assistants become more sophisticated and ubiquitous, how might they affect human memory, decision-making processes, and interpersonal relationships over time? As AI systems take on more complex cognitive tasks, how might this impact human skill development and labor markets in the long term?

Addressing these questions will require sustained, interdisciplinary research efforts and the development of new methodologies for studying the long-term societal impacts of technology. This could involve establishing long-term observatories for Human-AI Empowerment, developing predictive models for the societal impacts of AI, and creating forums for ongoing dialogue about the future trajectories of human-AI co-evolution.

6.10   THE CULTIVATION OF AI LITERACY

As AI systems become more integrated into various aspects of human life, the importance of widespread AI literacy cannot be overstated. Empowering individuals to understand, critically evaluate, and effectively engage with AI technologies is crucial for realizing the full potential of Human-AI Empowerment while mitigating potential risks [198].

AI literacy goes beyond basic digital literacy, requiring an understanding of AI concepts, capabilities, limitations, and ethical implications. Initiatives like Finland's "Elements of AI" course, which aims to educate 1% of European citizens about AI, represent important steps in this direction [241]. Future initiatives should focus on developing comprehensive AI education programs for various age groups and backgrounds, integrating AI literacy into school curricula, creating public awareness campaigns about AI, and promoting ongoing public discourse on the implications of AI technologies. This could involve developing interactive, hands-on learning experiences that allow individuals to experiment with AI technologies, creating citizen science projects that engage the public in AI research, and establishing community-based AI literacy programs.

In conclusion, the field of Human-AI Empowerment stands at an exciting and critical juncture. The potential for AI to enhance human capabilities, expand opportunities, and address global challenges is immense. However, realizing this potential requires careful navigation of complex ethical, social, and technical landscapes. As we move forward, it is crucial to maintain a balanced approach that harnesses the power of AI while prioritizing human values, agency, and well-being.

The future of Human-AI Empowerment will be shaped by our ability to foster true symbiosis between human and artificial intelligence, creating systems that amplify human strengths while compensating for our limitations. This will require ongoing interdisciplinary collaboration, a commitment to ethical and human-centered design, and the development of adaptive governance frameworks that can evolve alongside technological advancements.

Moreover, as we continue to push the boundaries of what is possible with AI, we must remain vigilant in our consideration of the broader implications of these technologies on human society and culture. The goal of Human-AI Empowerment should not merely be to create more efficient or capable systems, but to enhance human flourishing in its fullest sense – promoting creativity, compassion, and collective well-being.

The journey toward truly empowering Human-AI collaboration is just beginning. As researchers, practitioners, policymakers, and members of society, we all have a role to play in shaping this future. Approaching these challenges with creativity, critical thinking, and a steadfast commitment to human values allows us to work toward a future where AI serves as a powerful tool for human empowerment and the betterment of our world.

As we conclude this exploration, it is clear that the field of Human-AI Empowerment will continue to evolve rapidly in the coming years. The insights, frameworks, and case studies presented in this book provide a foundation for future research and development. However, many questions remain unanswered, and new challenges will undoubtedly emerge as AI technologies continue to advance. It is our hope that this work will inspire further inquiry, innovation, and collaboration in the pursuit of truly empowering Human-AI synergy.

The future of Human-AI Empowerment is not predetermined, but will be shaped by the choices we make today and in the years to come. Maintaining a focus on human-centered design, ethical considerations, and the broader societal implications of AI allows us to work toward a future where artificial intelligence serves as a powerful force for human flourishing and the advancement of our collective potential. The journey ahead is complex and challenging, but it is also filled with unprecedented opportunities for enhancing human capabilities and addressing the most pressing issues facing our world. As we embark on this journey, let us do so with wisdom, foresight, and an unwavering commitment to the betterment of humanity.

Bibliography

[1] Mohamed Abdalla and Moustafa Abdalla. Many top AI researchers get financial backing from big tech. *Wired*, 2020.

[2] Ashraf Abdul, Jo Vermeulen, Danding Wang, Brian Y Lim, and Mohan Kankanhalli. Trends and trajectories for explainable, accountable and intelligible systems: An HCI research agenda. *Proceedings of the 2018 CHI Conference on Human Factors in Computing Systems*, pages 1–18, 2018. ACM (Association for Computing Machinery).

[3] Daron Acemoglu. Technical change, inequality, and the labor market. *Journal of Economic Literature*, 40(1):7–72, 2002.

[4] Daron Acemoglu, David Autor, Jonathon Hazell, and Pascual Restrepo. AI and jobs: Evidence from online vacancies. *National Bureau of Economic Research*, 2020.

[5] Daron Acemoglu and Pascual Restrepo. Artificial intelligence, automation and work. *The economics of artificial intelligence: An agenda*, pages 197–236, 2018. ACM (Association for Computing Machinery).

[6] Marieke A Adriaanse, Johanna MF Van Oosten, Denise TD De Ridder, John BF De Wit, and Catharine Evers. Planning what not to eat: Ironic effects of implementation intentions negating unhealthy habits. *Personality and Social Psychology Bulletin*, 37(1):69–81, 2011.

[7] Ajay Agrawal, Joshua Gans, and Avi Goldfarb. *Prediction machines: The simple economics of artificial intelligence*. Harvard Business Press, 2018.

[8] Amina Ahmed, Lucas Pereira, and Kimberly Jane. Mixed methods research: Combining both qualitative and quantitative approaches. *ResearchGate*, 2024.

[9] Jorge A Ahumada, Eric Fegraus, Tanya Birch, Nicole Flores, Roland Kays, Timothy G O'Brien, Jonathan Palmer, Stephanie Schuttler, Jennifer Y Zhao, Walter Jetz, et al. Wildlife insights: A platform to maximize the potential of camera trap and other passive sensor wildlife data for the planet. *Environmental Conservation*, 47(1):1–6, 2020.

[10] Arash Ajoudani, Andrea Maria Zanchettin, Serena Ivaldi, Alin Albu-Schäffer, Kazuhiro Kosuge, and Oussama Khatib. Progress and prospects of the human–robot collaboration. *Autonomous Robots*, 42(5):957–975, 2018.

[11] Carlos Alario-Hoyos, Iria Estévez-Ayres, Mar Pérez-Sanagustjn, Carlos Delgado Kloos, and Carmen Fernández-Panadero. Understanding learners' motivation and learning strategies in moocs. *The International Review of Research in Open and Distributed Learning*, 18(3), 2017.

[12] Vincent Aleven, Elizabeth A McLaughlin, R Allen Glenn, and Kenneth R Koedinger. Instruction based on adaptive learning technologies. *Handbook of research on learning and instruction*, 2:522–560, 2016.

[13] Sabina Alkire. The capability approach and well-being measurement for public policy. In Matthew D Adler and Marc Fleurbaey, editors, *The Oxford handbook of well-being and public policy*, pages 615–644. Oxford University Press, 2015.

[14] Amtrak. Amtrak's virtual assistant Julie. `https://www.amtrak.com/about-julie-amtrak-virtual-travel-assistant`, 2018. Accessed: 2021-08-15.

[15] Anne Anastasi and Susana Urbina. *Psychological testing.* Prentice Hall/Pearson Education, 1997.

[16] Quentin Andre, Ziv Carmon, Klaus Wertenbroch, Christopher Chabris, Daniel Goldstein, and Joel Huber. Leveraging AI to enhance customer experience: Insights from the 2017 CRM/AI enabled consumer experience workshop. *Customer Needs and Solutions*, 5(1):1–11, 2018.

[17] Julia Angwin, Jeff Larson, Surya Mattu, and Lauren Kirchner. Machine bias. *ProPublica*, 23(2016):139–159, 2016.

[18] Gopala K Anumanchipalli, Josh Chartier, and Edward F Chang. Speech synthesis from neural decoding of spoken sentences. *Nature*, 568(7753):493–498, 2019.

[19] Syrian Archive. Syrian archive: Preserving digital evidence for justice and human rights. `https://syrianarchive.org/`, 2020. Accessed: 2021-08-15.

[20] Jay D Aronson, Vladimir Dubrovsky, Alison Reuther, Delia Xu, and Caitlin Tuholske. Video analytics for human rights monitoring. *Journal of Human Rights Practice*, 10(3):446–463, 2018.

[21] Artbreeder. Artbreeder: Collaborative AI art tool. `https://www.artbreeder.com/`, 2020. Accessed: 2021-08-15.

[22] Frank Arute, Kunal Arya, Ryan Babbush, Dave Bacon, Joseph C Bardin, Rami Barends, Rupak Biswas, Sergio Boixo, Fernando GSL Brandao, David A Buell, et al. Quantum supremacy using a programmable superconducting processor. *Nature*, 574(7779):505–510, 2019.

[23] United Nations Office at Geneva. Background on lethal autonomous weapons systems in the CCW. https://www.unog.ch/80256EE600585943, 2021. Accessed: 2021-08-15.

[24] Autodesk. Autodesk's generative design software. https://www.autodesk.com/solutions/generative-design, 2019. Accessed: 2021-08-15.

[25] David H Autor. Why are there still so many jobs? The history and future of workplace automation. *Journal of Economic Perspectives*, 29(3):3–30, 2015.

[26] Edmond Awad, Sohan Dsouza, Richard Kim, Jonathan Schulz, Joseph Henrich, Azim Shariff, Jean-François Bonnefon, and Iyad Rahwan. The moral machine experiment. *Nature*, 563(7729):59–64, 2018.

[27] Roger Azevedo and Dragan Gašević. Understanding the complex nature of self-regulatory processes in learning with computer-based learning environments: An introduction. *Computers in Human Behavior*, 78:59–63, 2018.

[28] Ryan S Baker. Stupid tutoring systems, intelligent humans. *International Journal of Artificial Intelligence in Education*, 26(2):600–614, 2016.

[29] Ryan SJD Baker and Paul Salvador Inventado. Educational data mining and learning analytics. *Learning analytics*, pages 61–75, 2014.

[30] Paul B Baltes. Longitudinal and cross-sectional sequences in the study of age and generation effects. *Human Development*, 11(3):145–171, 1979.

[31] Albert Bandura. Social cognitive theory of self-regulation. *Organizational Behavior and Human Decision Processes*, 50(2):248–287, 1991.

[32] Gabriel Barata, Sandra Gama, Joaquim Jorge, and Daniel Gonçalves. Improving participation and learning with gamification. *Proceedings of the First International Conference on Gameful Design, Research, and Applications*, pages 10–17, 2013.

[33] Marguerite Barry and Gavin Doherty. What we talk about when we talk about interactivity: Empowerment in public discourse. *New Media & Society*, 19(7):1052–1071, 2016.

[34] Roy F Baumeister, Ellen Bratslavsky, Mark Muraven, and Dianne M Tice. Ego depletion: Is the active self a limited resource? *Journal of Personality and Social Psychology*, 74(5):1252, 1998.

[35] Patricia Bazeley. *Integrating analyses in mixed methods research*. SAGE Publications Ltd, London, 2018.

[36] Jacob Biamonte, Peter Wittek, Nicola Pancotti, Patrick Rebentrost, Nathan Wiebe, and Seth Lloyd. Quantum machine learning. *Nature*, 549(7671):195–202, 2017.

[37] Mark Billinghurst, Adrian Clark, and Gun Lee. A survey of augmented reality. *Foundations and Trends in Human-Computer Interaction*, 8(2-3):73–272, 2015.

[38] Aravind BK. Navigating ethical considerations in industry-academia collaborations. https://www.linkedin.com/pulse/navigating-ethical-considerations-industry-academia-aravind-bk-rddyc, 2023. Accessed: 2025-04-14.

[39] Benjamin S Bloom. *Taxonomy of educational objectives: The classification of educational goals*. Longmans, Green, 1956.

[40] BlueDot. Bluedot: AI-powered infectious disease surveillance. https://bluedot.global/, 2020. Accessed: 2021-08-15.

[41] Iris Bohnet. *What works: Gender equality by design.* Harvard University Press, 2016.

[42] Niall Bolger, Angelina Davis, and Eshkol Rafaeli. Diary methods: Capturing life as it is lived. *Annual Review of Psychology*, 54(1):579–616, 2003.

[43] Stephen P Borgatti, Ajay Mehra, Daniel J Brass, and Giuseppe Labianca. Network analysis in the social sciences. *science*, 323(5916):892–895, 2009.

[44] Nick Bostrom. *Superintelligence: Paths, dangers, strategies.* Oxford University Press, 2014.

[45] Nick Bostrom and Julian Savulescu. Human enhancement ethics: The state of the debate. 2007.

[46] Sarah Bouhouita-Guermech, Patrick Gogognon, and Jean-Christophe Bélisle-Pipon. The ethical implications of using AI in qualitative research. *ResearchGate*, 2023.

[47] Elizabeth A Boyle, Thomas Hainey, Thomas M Connolly, Grant Gray, Jeffrey Earp, Michela Ott, Theodore Lim, Manuel Ninaus, Claudia Ribeiro, and João Pereira. An update to the systematic literature review of empirical evidence of the impacts and outcomes of computer games and serious games. *Computers & Education*, 94:178–192, 2016.

[48] Tom B Brown, Benjamin Mann, Nick Ryder, Melanie Subbiah, Jared Kaplan, Prafulla Dhariwal, Arvind Neelakantan, Pranav Shyam, Girish Sastry, Amanda Askell, et al. Language models are few-shot learners. *arXiv preprint arXiv:2005.14165*, 2020.

[49] Peter Brusilovsky and Christoph Peylo. Adaptive and intelligent web-based educational systems. *International Journal of Artificial Intelligence in Education*, 13(2-4):159–172, 2003.

[50] Erik Brynjolfsson and Andrew McAfee. *Machine, platform, crowd: Harnessing our digital future*. WW Norton & Company, 2017.

[51] Joy Buolamwini and Timnit Gebru. Gender shades: Intersectional accuracy disparities in commercial gender classification. *Conference on fairness, accountability and transparency*, pages 77–91, 2018.

[52] Lee D Butterfield, William A Borgen, Norman E Amundson, and Asa-Sophia T Maglio. Fifty years of the critical incident technique: 1954-2004 and beyond. *Qualitative Research*, 5(4):475–497, 2005.

[53] John T Cacioppo, Louis G Tassinary, and Gary Berntson. *Handbook of Psychophysiology*. Cambridge University Press, 2007.

[54] Ryan Calo. Robotics and the lessons of cyberlaw. *California Law Review*, 103:513, 2015.

[55] Ryan Calo. Artificial intelligence policy: A primer and roadmap. *UC Davis L. Rev.*, 51:399, 2017.

[56] Pablo Carbonell, Abdullah Gök, Philip Shapira, and Jean-Loup Faulon. Synthetic biology: Engineering complexity and refactoring cell capabilities. *Nature Reviews Molecular Cell Biology*, 20(9):563–575, 2019.

[57] Nicholas Carr. *The glass cage: How our computers are changing us*. WW Norton & Company, 2014.

[58] Shan Carter and Michael Nielsen. Using artificial intelligence to augment human intelligence. *Distill*, 2(12):e9, 2017.

[59] Corinne Cath, Sandra Wachter, Brent Mittelstadt, Mariarosaria Taddeo, and Luciano Floridi. Governing artificial intelligence: Ethical, legal and technical opportunities and challenges. *Philosophical Transactions of the Royal Society A: Mathematical, Physical and Engineering Sciences*, 376(2133):20180080, 2018.

[60] Alberto J Cañas and Joseph D Novak. Concept maps: Theory, methodology, technology. *Proceedings of the First International Conference on Concept Mapping*, 1:2–10, 2005.

[61] Ling Cen, Dymitr Ruta, Jason Ng, and Raad Al-Hakim. Matching and generalization: A new algorithm for educational data mining. *Computers & Education*, 94:114–127, 2016.

[62] Kathy Charmaz. *Constructing grounded theory.* SAGE Publications, 2014.

[63] Kuan-Chung Chen and Syh-Jong Jang. Self-determination theory in game-based learning. *Simulation & Gaming*, 46(2):159–196, 2015.

[64] Tsan-Ming Choi and Shan Guo. Artificial intelligence for the future of work: A systematic literature review. *International Journal of Information Management*, 55:102182, 2020.

[65] Andy Clark. *Natural-born cyborgs: Minds, technologies, and the future of human intelligence.* Oxford University Press, 2003.

[66] Ruth Berins Collier and David Collier. *Shaping the political arena: Critical junctures, the labor movement, and regime dynamics in Latin America.* Princeton University Press, 1991.

[67] Simon Colton, Alison Pease, and Rob Saunders. Issues of authenticity in autonomously creative systems. *Proceedings of the 9th International Conference on Computational Creativity*, pages 272–279, 2018.

[68] European Commission. Proposal for a regulation laying down harmonised rules on artificial intelligence. https://digital-strategy.ec.europa.eu/en/

library/proposal-regulation-laying-down-
harmonised-rules-artificial-intelligence, 2021.
Accessed: 2021-08-15.

[69] John W Creswell and J David Creswell. *Research design: Qualitative, quantitative, and mixed methods approaches.* SAGE Publications, 2017.

[70] John W Creswell and Vicki L Plano Clark. *Designing and Conducting Mixed Methods Research.* SAGE Publications, 2017.

[71] Mihaly Csikszentmihalyi. *Flow: The psychology of optimal experience.* Harper & Row, 1990.

[72] Hengchen Dai, Katherine L Milkman, and Jason Riis. The fresh start effect: Temporal landmarks motivate aspirational behavior. *Management Science*, 60(10):2563–2582, 2014.

[73] Paul R Daugherty and H James Wilson. *Human+ machine: Reimagining work in the age of AI.* Harvard Business Press, 2018.

[74] Maartje MA de Graaf, Somaya Ben Allouch, and Tineke Klamer. Designing ethical social robots—a longitudinal field study with older adults. *Frontiers in Robotics and AI*, 7:1–13, 2020.

[75] Ewart J de Visser, Richard Pak, and Tyler H Shaw. Toward a theory of longitudinal trust calibration in human–robot teams. *International Journal of Social Robotics*, 12(2):459–478, 2020.

[76] Stanislas Dehaene. *How we learn: Why brains learn better than any machine... for now.* Viking, 2020.

[77] Norman K Denzin and Yvonna S Lincoln. *The SAGE handbook of qualitative research.* SAGE, 2011.

[78] Sebastian Deterding, Dan Dixon, Rilla Khaled, and Lennart Nacke. From game design elements to gamefulness: defining gamification. *Proceedings of the 15th International Academic MindTrek Conference: Envisioning Future Media Environments*, pages 9–15, 2011.

[79] Christo Dichev and Darina Dicheva. Gamification in education: A systematic mapping study. *Educational Technology & Society*, 20(3):14–31, 2017.

[80] Donna Dickson and Rand Nusair. Artificial intelligence in talent acquisition: a review of AI-applications used in recruitment and selection. *Strategic HR Review*, 18(5):215–221, 2019.

[81] Joshua J Diehl, Lauren M Schmitt, Michael Villano, and Charles R Crowell. The clinical use of robots for individuals with autism spectrum disorders: A critical review. *Research in autism spectrum disorders*, 6(1):249–262, 2012.

[82] Pierre Dillenbourg. What do you mean by collaborative learning? *Collaborative-Learning: Cognitive and Computational Approaches*, 1:1–15, 2009.

[83] Angelika Dimoka. How to conduct a functional magnetic resonance (FMRI) study in social science research. *MIS Quarterly*, 811–840, 2012.

[84] Carl DiSalvo, Illah Nourbakhsh, David Holstius, Ayşe Akin, and Marti Louw. The neighborhood networks project: A case study of technology-supported community building. In *Proceedings of the SIGCHI Conference on Human Factors in Computing Systems*, pages 51–60, 2010.

[85] Finale Doshi-Velez and Been Kim. Towards a rigorous science of interpretable machine learning. *arXiv preprint arXiv:1702.08608*, 2017.

[86] Carol S Dweck. *Mindset: The new psychology of success.* Random House Digital, Inc., 2006.

[87] Eneza Education. Eneza education: AI-powered mobile learning platform. `https://enezaeducation.com/`, 2019. Accessed: 2021-08-15.

[88] Glen H Elder Jr. The life course as developmental theory. *Child development*, 69(1):1–12, 1998.

[89] Ahmed Elgammal, Bingchen Liu, Mohamed Elhoseiny, and Marian Mazzone. Can: Creative adversarial networks, generating "art" by learning about styles and deviating from style norms. *arXiv preprint arXiv:1706.07068*, 2017.

[90] Madeleine Clare Elish. Moral crumple zones: Cautionary tales in human-robot interaction. *Engaging Science, Technology, and Society*, 5:40–60, 2019.

[91] Douglas C Engelbart. Augmenting human intellect: A conceptual framework. *Summary Report AFOSR-3223 under Contract AF 49 (638)-1024, SRI Project 3578 for Air Force Office of Scientific Research*, 1962.

[92] Ziv Epstein, Aaron Hertzmann, Laura Herman, Robert Mahari, Morgan R Frank, Matthew Groh, Hope Schroeder, Amy Smith, Memo Akten, Jessica Fjeld, Hany Farid, Neil Leach, Alex Pentland, and Olga Russakovsky. Art and the science of generative AI: A deeper dive. *arXiv preprint arXiv:2306.04141*, 2023.

[93] Andre Esteva, Brett Kuprel, Roberto A Novoa, Justin Ko, Susan M Swetter, Helen M Blau, and Sebastian Thrun. Dermatologist-level classification of skin cancer with deep neural networks. *Nature*, 542(7639):115–118, 2017.

[94] Amitai Etzioni and Oren Etzioni. Incorporating ethics into artificial intelligence. *The Journal of Ethics*, 21(4):403–418, 2019.

[95] Virginia Eubanks. *Automating inequality: How high-tech tools profile, police, and punish the poor.* St. Martin's Press, 2018.

[96] Jerry Alan Fails and Dan R Olsen Jr. Interactive machine learning. In *Proceedings of the 8th International Conference on Intelligent User Interfaces*, pages 39–45. ACM, 2003.

[97] Fei Fang, Thanh H Nguyen, Rob Pickles, Wai Y Lam, Gopalasamy R Clements, Bo An, Amandeep Singh, Milind Tambe, and Andrew Lemieux. Deploying paws: Field optimization of the protection assistant for wildlife security. *Twenty-Ninth IAAI Conference*, 2017.

[98] Andy Field. *Discovering statistics using IBM SPSS statistics.* SAGE, 2013.

[99] Leah Findlater and Jacob Wobbrock. Personalized input: Improving ten-finger touchscreen typing through automatic adaptation. *Proceedings of the SIGCHI Conference on Human Factors in Computing Systems*, pages 1817–1826, 2009.

[100] Gerhard Fischer. Lifelong learning—more than training. *Journal of Interactive Learning Research*, 11(3):265–294, 2000.

[101] Kathleen Kara Fitzpatrick, Alison Darcy, and Molly Vierhile. Delivering cognitive behavior therapy to young adults with symptoms of depression and anxiety using a fully automated conversational agent (woebot): A randomized controlled trial. *JMIR Mental Health*, 4(2):e19, 2017.

[102] Stitch Fix. Stitch fix's hybrid design process. `https://algorithms-tour.stitchfix.com/`, 2018. Accessed: 2021-08-15.

[103] John C Flanagan. The critical incident technique. *Psychological Bulletin*, 51(4):327, 1954.

[104] Sharlene N Flesher, John E Downey, Jeffrey M Weiss, Cory L Hughes, Angelica J Herrera, Elizabeth C Tyler-Kabara, Michael L Boninger, Jennifer L Collinger, and Robert A Gaunt. A brain-computer interface that evokes tactile sensations improves robotic arm control. *Science*, 372(6544):831–836, 2021.

[105] Luciano Floridi. AI and its new winter: From myths to realities. *Philosophy & Technology*, 33(1):1–3, 2020.

[106] Luciano Floridi, Josh Cowls, Monica Beltrametti, Raja Chatila, Patrice Chazerand, Virginia Dignum, Christoph Luetge, Robert Madelin, Ugo Pagallo, Francesca Rossi, et al. Ai4people—an ethical framework for a good AI society: Opportunities, risks, principles, and recommendations. *Minds and Machines*, 28(4):689–707, 2018.

[107] Peter W Foltz, Lynn A Streeter, Karen E Lochbaum, and Thomas K Landauer. The sources of text difficulty: Evidence from automated essay scoring. *Handbook of automated essay evaluation: Current applications and new directions*, pages 196–219, 2013.

[108] U.S. Food and Drug Administration. Artificial intelligence/machine learning (AI/ML)-based software as a medical device (SAMD) action plan. https://www.fda.gov/medical-devices/software-medical-device-samd/artificial-intelligence-and-machine-learning-software-medical-device, 2021. Accessed: 2021-08-15.

[109] United Nations Institute for Training and Research. Unitar's ZAC (zonal automated classifier). https://unitar.org/sustainable-development-goals/satellite-analysis-and-applied-research, 2019. Accessed: 2021-08-15.

[110] Martin Ford. *Rise of the robots: Technology and the threat of a jobless future.* Basic Books, 2015.

[111] Floyd J Fowler Jr. *Survey research methods.* SAGE Publications, 2013.

[112] Harry G Frankfurt. Freedom of the will and the concept of a person. *The Journal of Philosophy*, 68(1):5–20, 1971.

[113] Jonas Frich, Michael Mose Biskjaer, and Peter Dalsgaard. Mapping the landscape of human-AI interaction in creative practice. *Proceedings of the 2019 on Creativity and Cognition*, pages 389–401, 2019.

[114] Luke K Fryer, Kaori Nakao, and Amy Thompson. Chatbots in education: A critical review. *Computer Assisted Language Learning*, 32(8):1006–1033, 2019.

[115] Kentaro Fujita, Peter M Gollwitzer, and Gabriele Oettingen. Mindset and pre-conscious open-mindedness to incidental information. *Journal of Experimental Social Psychology*, 44(5):1260–1266, 2008.

[116] Fujitsu. Fujitsu's AI-powered worker safety system. https://www.fujitsu.com/global/about/resources/news/press-releases/2019/0619-01.html, 2019. Accessed: 2021-08-15.

[117] Krzysztof Z Gajos, Mary Czerwinski, Desney S Tan, and Daniel S Weld. Automatically generating personalized user interfaces with supple. *Artificial Intelligence*, 170(10):910–950, 2006.

[118] Clare Garvie, Alvaro Bedoya, and Jonathan Frankle. The perpetual line-up: Unregulated police face recognition in america. *Georgetown Law Center on Privacy & Technology*, 2016.

[119] James Paul Gee. *An introduction to discourse analysis: Theory and method.* Routledge, 2014.

[120] Roxana Girju. Understanding lived experience: Bridging artificial intelligence and natural language processing with humanities and social sciences. In *IOP Conference Series: Materials Science and Engineering*, volume 1292, page 012020. IOP Publishing, 2023.

[121] Barney G Glaser and Anselm L Strauss. *The discovery of grounded theory: Strategies for qualitative research*. Aldine Publishing Company, 1967.

[122] Robert Godwin-Jones. Smartphones and language learning. *Language Learning & Technology*, 21(2):3–17, 2017.

[123] Ashok K Goel and Lalith Polepeddi. AI-powered learning: Making education accessible, affordable, and achievable. *arXiv preprint arXiv:2006.01908*, 2020.

[124] Peter M Gollwitzer. Implementation intentions: Strong effects of simple plans. *American Psychologist*, 54(7):493, 1999.

[125] Peter M Gollwitzer and Paschal Sheeran. Implementation intentions and goal achievement: A meta-analysis of effects and processes. *Advances in Experimental Social Psychology*, 38:69–119, 2006.

[126] Google. Live transcribe: Speech to text, sound notifications. https://www.android.com/accessibility/live-transcribe/, 2019. Accessed: 2021-08-15.

[127] Katja Grace, John Salvatier, Allan Dafoe, Baobao Zhang, and Owain Evans. When will AI exceed human performance? evidence from AI experts. *Journal of Artificial Intelligence Research*, 62:729–754, 2018.

[128] Jennifer C Greene, Valerie J Caracelli, and Wendy F Graham. Toward a conceptual framework for mixed-method evaluation designs. *Educational Evaluation and Policy Analysis*, 11(3):255–274, 1989.

[129] David Gunning and David W Aha. Darpa's explainable artificial intelligence program. *AI Magazine*, 40(2):44–58, 2019.

[130] Kristin Gustavson, Tilmann von Soest, Evalill Karevold, and Espen RØysamb. Attrition and generalizability in longitudinal studies: Findings from a 15-year population-based study and a Monte Carlo simulation study. *BMC Public Health*, 12(1):1–11, 2012.

[131] Kristin Gustavson, Tilmann von Soest, Evalill Karevold, and Espen RØysamb. Attrition and generalizability in longitudinal studies: Findings from a 15-year population-based study and a Monte Carlo simulation study. *BMC Public Health*, 12(1):918, 2012.

[132] Leah Hamilton, Desha Elliott, Aaron Quick, Simone Smith, and Victoria Choplin. Exploring the use of AI in qualitative analysis: A comparative study of guaranteed income data. *International Journal of Qualitative Methods*, 22:1–13, 2023.

[133] Martyn Hammersley and Paul Atkinson. *Ethnography: Principles in practice*. Routledge, 2007.

[134] Adam Hampshire, Roger R Highfield, Beth L Parkin, and Adrian M Owen. Fractionating human intelligence. *Neuron*, 76(6):1225–1237, 2012.

[135] Douglas Harper. Talking about pictures: A case for photo elicitation. *Visual Studies*, 17(1):13–26, 2002.

[136] Erik Harpstead, Brad A Myers, and Vincent Aleven. In search of learning: facilitating data analysis in educational games. In *Proceedings of the SIGCHI Conference on Human Factors in Computing Systems*, pages 79–88, 2013.

[137] Demis Hassabis, Dharshan Kumaran, Christopher Summerfield, and Matthew Botvinick. Neuroscience-inspired artificial intelligence. *Neuron*, 95(2):245–258, 2017.

[138] Marc Hassenzahl. *Experience design: Technology for all the right reasons.* Morgan & Claypool Publishers, 2010.

[139] Dirk Helbing, Bruno S Frey, Gerd Gigerenzer, Ernst Hafen, Michael Hagner, Yvonne Hofstetter, Jeroen Van Den Hoven, Roberto V Zicari, and Andrej Zwitter. Will democracy survive big data and artificial intelligence? *Towards digital enlightenment*, pages 73–98, 2019.

[140] Dorien Herremans, Ching-Hua Chuan, and Elaine Chew. Functional scaffolding for composing additional musical voices. *Computer Music Journal*, 41(2):79–95, 2017.

[141] Mireille Hildebrandt. Privacy as protection of the incomputable self: From agnostic to agonistic machine learning. *Theoretical Inquiries in Law*, 20(1):83–121, 2019.

[142] Kashmir Hill. The secretive company that might end privacy as we know it. *The New York Times*, 18, 2020.

[143] Wayne Holmes, Maya Bialik, and Charles Fadel. *Artificial intelligence in education promises and implications for teaching and learning.* Center for Curriculum Redesign, 2019.

[144] Andreas Holzinger. Towards interactive machine learning (IML): Applying artificial intelligence to enable knowledge creation processes in explorative research practices. *Proceedings of the 2nd International Conference on Computer Science and Application Engineering*, pages 1–5, 2017.

[145] Kristina Höök. Steps to a theory of somaesthetic design. *ACM Transactions on Computer-Human Interaction (TOCHI)*, 25(6):1–30, 2018.

[146] Michael C Horowitz. Artificial intelligence and international security. *Center for a New American Security*, 2018.

[147] Matthew B Hoy. Alexa, Siri, Cortana, and more: An introduction to voice assistants. *Medical Reference Services Quarterly*, 37(1):81–88, 2018.

[148] Edwin Hutchins. *Cognition in the wild*. MIT Press, 1995.

[149] IBM. IBM's your learning platform. https://www.ibm.com/training/, 2018. Accessed: 2021-08-15.

[150] IBM. AI fairness 360: An open source toolkit to detect and mitigate unwanted bias in machine learning models. https://aif360.mybluemix.net/, 2020. Accessed: 2021-08-15.

[151] Don Ihde. *Technology and the lifeworld: From garden to earth*. Indiana University Press, 1990.

[152] Masumi Iida, Patrick E Shrout, Jean-Philippe Laurenceau, and Niall Bolger. Using diary methods in psychological research. *APA handbook of research methods in psychology*, 1:277–305, 2012. American Psychological Association.

[153] Becky Inkster, Shubhankar Sarda, and Vinod Subramanian. An empathy-driven, conversational artificial intelligence agent (WYSA) for digital mental well-being: Real-world data evaluation mixed-methods study. *JMIR mHealth and uHealth*, 6(11):e12106, 2018.

[154] Michael Inzlicht, Brandon J Schmeichel, and C Neil Macrae. What is ego depletion? Toward a mechanistic revision of the resource model of self-control. *Perspectives on Psychological Science*, 11(1):45–58, 2016.

[155] John PA Ioannidis. Why most published research findings are false. *PLoS Medicine*, 2(8):e124, 2005.

[156] Mark Israel. *Research ethics and integrity for social scientists: Beyond regulatory compliance*. SAGE, 2014.

[157] Joichi Ito. The next step for AI is human-centered AI. *Wired*, 2018.

[158] Giulio Jacucci, Ann Morrison, Gabriela T Richard, Maaike Kleinsmann, and Neville A Stanton. Designing for effective collaboration in tech-enhanced environments. *Proceedings of the 2014 Conference on Designing Interactive Systems*, pages 11–14, 2014.

[159] Ravin Jesuthasan and John W Boudreau. Reinventing jobs: A 4-step approach for applying automation to work. *Harvard Business Review Press*, 2018.

[160] Anna Jobin, Marcello Ienca, and Effy Vayena. Artificial intelligence: the global landscape of ethics guidelines. *Nature Machine Intelligence*, 1(9):389–399, 2019.

[161] Oliver P John and Sanjay Srivastava. The big five trait taxonomy: History, measurement, and theoretical perspectives. *Handbook of personality: Theory and research*, 2(1999):102–138, 1999.

[162] John Jumper, Richard Evans, Alexander Pritzel, Tim Green, Michael Figurnov, Olaf Ronneberger, Kathryn Tunyasuvunakool, Russ Bates, Augustin Žjdek, Anna Potapenko, et al. Highly accurate protein structure prediction with alphafold. *Nature*, 596(7873):583–589, 2021.

[163] Daniel Kahneman and Amos Tversky. Prospect theory: An analysis of decision under risk. *Econometrica*, 47(2):263–291, 1979.

[164] Georgios A Kaissis, Marcus R Makowski, Daniel Rückert, and Rickmer F Braren. Secure, privacy-preserving and federated machine learning in medical imaging. *Nature Machine Intelligence*, 2(6):305–311, 2020.

[165] Ece Kamar. Directions in hybrid intelligence: Complementing AI systems with human intelligence. *IJCAI*, pages 4070–4073, 2016.

[166] Maria Kandaurova and Daniel A. Skog. Initiating and expanding data network effects: A longitudinal case study of generativity in the evolution of an AI platform. In *Proceedings of the 57th Hawaii International Conference on System Sciences*, pages 6250–6259. University of Hawai'i at Manoa, 2024.

[167] Anna Kantosalo, Hannu Toivonen, Ping Xiao, and Jukka M Toivanen. Modes for creative human-computer collaboration: Alternating and task-divided co-creativity. *Proceedings of the 7th International Conference on Computational Creativity*, 2014.

[168] Kamal Kapoor, Arnim Wiek, Roberto E Galang, Chelsea Davis, and Vicki Prewitt. Just enough to be dangerous: Creativity and the role of knowledge in generating ideas. *Creativity Research Journal*, 27(2):131–142, 2015.

[169] Fabre, Émilie, Katie Seaborn, Shuta Koiwai, Mizuki Watanabe, and Paul Riesch. *More-than-Human Storytelling: Designing Longitudinal Narrative Engagements with Generative AI*. In Proceedings of the Extended Abstracts of the CHI Conference on Human Factors in Computing Systems, pages 1–10, 2025.

[170] Evangelos Karapanos, John Zimmerman, Jodi Forlizzi, and Jean-Bernard Martens. User experience over time: An initial framework. *Proceedings of the SIGCHI Conference on Human Factors in Computing Systems*, pages 729–738, 2009.

[171] Michael Katell, Meg Young, Dharma Dailey, Bernease Herman, Vivian Guetler, Aaron Tam, Corinne Bintz, Daniella Raz, and P M Krafft. Toward situated interventions for algorithmic equity: Lessons from the field. *Proceedings of the 2020 Conference on Fairness, Accountability, and Transparency*, pages 45–55, 2020.

[172] Florian Keusch, Christopher Antoun, Mick P Couper, Frauke Kreuter, and Lars Lyberg. Mobile technologies for conducting, augmenting and potentially replacing surveys: Executive summary of the aapor task force on emerging technologies in public opinion research. *Public Opinion Quarterly*, 79(4):801–819, 2015.

[173] Jan Kietzmann, Jeannette Paschen, and Emily Treen. Artificial intelligence in advertising: How marketers can leverage artificial intelligence along the consumer journey. *Journal of Advertising Research*, 59(3):263–267, 2019.

[174] John KC Kingston. Artificial intelligence and legal liability. *arXiv preprint arXiv:1802.07782*, 2018.

[175] Hiroaki Kitano. Artificial intelligence to win the nobel prize and beyond: Creating the engine for scientific discovery. *AI magazine*, 37(1):39–49, 2016.

[176] Branko Kolarevic. From mass customisation to design "democratisation." *Architectural Design*, 85(6):48–53, 2015.

[177] Anton Korinek and Joseph E Stiglitz. Artificial intelligence and its implications for income distribution and unemployment. *National Bureau of Economic Research*, 2017.

[178] Michal Kosinski, David Stillwell, and Thore Graepel. Private traits and attributes are predictable from digital records of human behavior. *Proceedings of the national academy of sciences*, 110(15):5802–5805, 2013.

[179] Richard A Krueger and Mary Anne Casey. *Focus groups: A practical guide for applied research*. SAGE Publications, 2014.

[180] Steinar Kvale. *Doing interviews*. SAGE, 2008.

[181] Joseph C Kvedar, Alexander L Fogel, Eric Elenko, and Daphne Zohar. Digital medicine's march on chronic disease. *Nature Biotechnology*, 37(8):916–924, 2019.

[182] Stanford Law School Legal Design Lab. Legal design lab: Justicebot. https://law.stanford.edu/organizations/pages/legal-design-lab/, 2019. Accessed: 2021-08-15.

[183] Brenden M Lake, Tomer D Ullman, Joshua B Tenenbaum, and Samuel J Gershman. Building machines that learn and think like people. *Behavioral and Brain Sciences*, 40:e253, 2017.

[184] Brenden M Lake, Tomer D Ullman, Joshua B Tenenbaum, and Samuel J Gershman. Building machines that learn and think like people. *Behavioral and brain sciences*, 41, 2018.

[185] Phillippa Lally, Cornelia HM Van Jaarsveld, Henry WW Potts, and Jane Wardle. How are habits formed: Modelling habit formation in the real world. *European Journal of Social Psychology*, 40(6):998–1009, 2010.

[186] Richard N Landers. Developing a theory of gamified learning: Linking serious games and gamification of learning. *Simulation & Gaming*, 45(6):752–768, 2014.

[187] Gary P Latham and Craig C Pinder. What should we do about motivation theory? Six recommendations for the twenty-first century. *Academy of Management Review*, 29(3):388–403, 2004.

[188] David Lazer, Alex Pentland, Lada Adamic, Sinan Aral, Albert-Laszlo Barabasi, Devon Brewer, Nicholas Christakis, Noshir Contractor, James Fowler, Myron Gutmann, et al. Computational social science. *Science*, 323(5915):721–723, 2009.

[189] Mikhail A Lebedev and Miguel AL Nicolelis. Brain-machine interfaces: from basic science to neuroprostheses and neurorehabilitation. *Physiological reviews*, 2017.

[190] Jay Lee, Moslem Azamfar, and Jaskaran Singh. Prognostics and health management of manufacturing systems. *Procedia Manufacturing*, 34:10–19, 2019.

[191] Kai-Fu Lee. *AI superpowers: China, Silicon Valley, and the new world order*. Houghton Mifflin Harcourt, 2018.

[192] Ian Li, Anind K Dey, and Jodi Forlizzi. Personal informatics and context: Using context to reveal factors that affect behavior. *Journal of Ambient Intelligence and Smart Environments*, 3(1):51–67, 2011.

[193] Konstantinos G Liakos, Patrizia Busato, Dimitrios Moshou, Simon Pearson, and Dionysis Bochtis. Machine learning in agriculture: A review. *Sensors*, 18(8):2674, 2018.

[194] JCR Licklider. Man-computer symbiosis. *IRE Transactions on Human Factors in Electronics*, 1(1):4–11, 1960.

[195] Tanya Linden, Kewei Yuan, and Antonette Mendoza. Phenomenological study of generative AI in higher education as perceived by academics. In *Proceedings of the ISCAP Conference*, Baltimore, MD, 2024.

[196] Yun Liu, Timo Kohlberger, Mohammad Norouzi, George E Dahl, Jenny L Smith, Arash Mohtashamian, Nathan Olson, Lily H Peng, Jason D Hipp, and Martin C Stumpe. Artificial intelligence in cancer imaging and diagnosis: opportunities and challenges. *CA: A Cancer Journal for Clinicians*, 68(6):452–466, 2018.

[197] Edwin A Locke and Gary P Latham. Building a practically useful theory of goal setting and task motivation: A 35-year odyssey. *American Psychologist*, 57(9):705, 2002.

[198] Dianna Long and Brian Magerko. Implications of artificial intelligence for conceptualizing AI literacy: A scoping review. *AI & SOCIETY*, pages 1–26, 2020.

[199] Todd Lubart. How can computers be partners in the creative process: classification and commentary on the special issue. *International Journal of Human-Computer Studies*, 63(4-5):365–369, 2005.

[200] Leanne Luce. Artificial intelligence for fashion: How AI is revolutionizing the fashion industry. *Apress*, 2019.

[201] Hannele Lukkarinen. Methodological triangulation showed the poorest quality of life in the youngest people following treatment of coronary artery disease: A longitudinal study. *International Journal of Nursing Studies*, 42(6):619–627, 2005.

[202] Rohit Madan and Mona Ashok. Making sense of AI benefits: A mixed-method study in Canadian public administration. *Information Systems Frontiers*, 2024.

[203] Viswakanth Makutam, Sai Yashashwini Achanti, and Marjan Doostan. Integration of artificial intelligence in adaptive trial designs: Enhancing efficiency and patient-centric outcomes. *International Journal of Advanced Research*, 12(8):205–215, 2024.

[204] Thomas W Malone. *Superminds: The surprising power of people and computers thinking together*. Little, Brown, 2018.

[205] Thomas W Malone and Michael S Bernstein. Collective intelligence and group performance. *Current Directions in Psychological Science*, 24(6):420–424, 2015.

[206] Lev Manovich. AI aesthetics. *Strelka Press*, 2018.

[207] Francesco Marconi. *Newsmakers: Artificial intelligence and the future of journalism*. Columbia University Press, 2019.

[208] Gary Marcus. The next decade in AI: four steps towards robust artificial intelligence. *arXiv preprint arXiv:2002.06177*, 2020.

[209] Theresa M Marteau, David Ogilvie, Martin Roland, Marc Suhrcke, and Michael P Kelly. Judging nudging: Can nudging improve population health? *BMJ*, 342, 2011.

[210] Becky McCall. Covid-19 and artificial intelligence: Protecting health-care workers and curbing the spread. *The Lancet Digital Health*, 2(4):e166–e167, 2020.

[211] John O McGinnis and Russell G Pearce. The great disruption: How machine intelligence will transform the role of lawyers in the delivery of legal services. *Fordham L. Rev.*, 82:3041, 2019.

[212] Robert McKee. *Story: Substance, structure, style, and the principles of screenwriting*. Methuen, 2019.

[213] Medtronic. Sugar.iq intelligent diabetes assistant. https://www.medtronicdiabetes.com/products/sugar-iq-diabetes-assistant, 2018. Accessed: 2021-08-15.

[214] Ninareh Mehrabi, Fred Morstatter, Nripsuta Saxena, Kristina Lerman, and Aram Galstyan. A survey on bias and fairness in machine learning. *arXiv preprint arXiv:1908.09635*, 2019.

[215] Anneloes Meijnders, Cees Midden, Anna Olofsson, Susanna Öhman, Jörg Matthes, Olha Bondarenko, and Jan Gutteling. Creating inspiring stories: Narrative approach to behavior change in energy consumption. *Journal of Environmental Psychology*, 57:95–107, 2018.

[216] Jeanne C Meister and Robert H Brown. The future of work: The intersection of artificial intelligence and human resources. *Human Resource Development International*, 23(5):535–545, 2020.

[217] Scott Menard. *Longitudinal research*. SAGE, 2002.

[218] Microsoft. Farmbeats: AI & IoT for agriculture. https://www.microsoft.com/en-us/research/project/farmbeats-iot-agriculture/, 2019. Accessed: 2021-08-15.

[219] Microsoft. Hololens 2: Mixed reality technology for business. https://www.microsoft.com/en-us/hololens, 2021. Accessed: 2021-08-15.

[220] Tim Miller. Explanation in artificial intelligence: Insights from the social sciences. *Artificial Intelligence*, 267:1–38, 2019.

[221] Antonija Mitrovic. Thirty years of intelligent tutoring systems: What have we learned? *International Journal of Artificial Intelligence in Education*, 25(2):247–266, 2015.

[222] Brent Daniel Mittelstadt, Patrick Allo, Mariarosaria Taddeo, Sandra Wachter, and Luciano Floridi. The ethics of algorithms: Mapping the debate. *Big Data & Society*, 3(2):2053951716679679, 2016.

[223] David L Morgan. Focus groups. *Annual Review of Sociology*, 22(1):129–152, 1996.

[224] Stephen L Morgan and Christopher Winship. *Counterfactuals and causal inference: Methods and principles for social research*. Cambridge University Press, 2007.

[225] Franz Moritz, Ivan Soltesz, Ran Gan, Balint Varkuti, Sandro M Krieg, Karl-Heinz Nenning, Georg Langs, Christopher Stipe, Evan S Lutkenhoff, Martin M Monti, et al. Predicting the spatiotemporal diversity of seizure propagation and termination in human focal epilepsy. *Nature Communications*, 10(1):1–13, 2019.

[226] Jesper Mortensen and Preben Lund. The business value of artificial intelligence. *IESE Insight*, 31:33–39, 2016.

[227] Geoff Mulgan. Artificial intelligence and collective intelligence: The emergence of a new field. *AI & SOCIETY*, 33(4):631–632, 2018.

[228] Stefan Munzer, Hubert Zimmmer, Karin Schwaibold, and Jorg Baus. Does using a GPS impair our ability to navigate? *Journal of Environmental Psychology*, 32(2):197–209, 2012.

[229] Wilhelmina Nekoto, Vukosi Marivate, Tshinondiwa Matsila, Timi Fasubaa, Tajudeen Fagbohungbe, Solomon Oluwole Akinola, Shamsuddeen Hassan Muhammad, Salomon Kabongo, Salomey Osei, Freshia Sackey, et al. Participatory research for low-resourced machine translation: A case study in african languages. *arXiv preprint arXiv:2010.02353*, 2020.

[230] Lokesh P Nijhawan, Manthan D Janodia, BS Muddukrishna, Karkala M Bhat, Kishan L Bairy, Nayanabhirama Udupa, and Prashant B Musmade. Informed consent: Issues and challenges. *Journal of Advanced Pharmaceutical Technology & Research*, 4(3):134, 2013.

[231] Helen Nissenbaum. *Privacy in context: Technology, policy, and the integrity of social life*. Stanford University Press, 2009.

[232] Safiya U. Noble. *Algorithms of Oppression: How search engines reinforce racism*. NYU Press, 2018.

[233] John C Norcross, Paul M Krebs, and James O Prochaska. Stages of change. *Journal of Clinical Psychology*, 67(2):143–154, 2011.

[234] Donald A Norman. *The design of everyday things: Revised and expanded edition*. Basic Books, 2013.

[235] Martha C Nussbaum. *Creating capabilities*. Harvard University Press, 2011.

[236] Martha C Nussbaum. *Creating capabilities: The human development approach.* Harvard University Press, 2011.

[237] Mercy Nyamewaa Asiedu, Iskandar Haykel, Awa Dieng, Kerrie Kauer, Tousif Ahmed, Florence Ofori, Charisma Chan, Stephen Pfohl, Negar Rostamzadeh, and Katherine Heller. Nteasee: A mixed methods study of expert and general population perspectives on deploying AI for health in african countries. *arXiv preprint arXiv:2409.12197,* 2024.

[238] Benjamin D Nye. Intelligent tutoring systems by and for the developing world: A review of trends and approaches for educational technology in a global context. *International Journal of Artificial Intelligence in Education,* 25(2):177–203, 2015.

[239] OECD. OECD principles on artificial intelligence. https://www.oecd.org/going-digital/ai/principles/, 2019. Accessed: 2021-08-15.

[240] Government of Canada. Directive on automated decision-making. https://www.tbs-sct.gc.ca/pol/doc-eng.aspx?id=32592, 2019. Accessed: 2021-08-15.

[241] University of Helsinki and Reaktor. Elements of AI. https://www.elementsofai.com/, 2021. Accessed: 2021-08-15.

[242] National Institute of Standards and Technology. Privacy-enhancing cryptography. https://www.nist.gov/programs-projects/privacy-enhancing-cryptography, 2020. Accessed: 2021-08-15.

[243] Partnership on AI. Partnership on AI. https://www.partnershiponai.org/, 2021. Accessed: 2021-08-15.

[244] IEEE Global Initiative on Ethics of Autonomous and Intelligent Systems. Ethically aligned design. https://ethicsinaction.ieee.org/, 2019. Accessed: 2021-08-15.

[245] Cathy O'Neil. *Weapons of math destruction: How big data increases inequality and threatens democracy.* Crown, 2016.

[246] Anthony J. Onwuegbuzie and Nancy L. Leech. Linking research questions to mixed methods data analysis procedures. *The Qualitative Report*, 11(3):474–498, 2006.

[247] Yoshinori Oyama, Emmanuel Manalo, and Yoshihide Nakatani. The hemingway effect: How failing to finish a task can have a positive effect on motivation. *Thinking Skills and Creativity*, 30:7–18, 2018.

[248] Fred Paas, Alexander Renkl, and John Sweller. Cognitive load theory and instructional design: Recent developments. *Educational Psychologist*, 38(1):1–4, 2003.

[249] John F Pane, Beth Ann Griffin, Daniel F McCaffrey, and Rita Karam. An effectiveness trial of the cognitive tutor algebra i program. *Journal of Research on Educational Effectiveness*, 7(4):396–421, 2014.

[250] Eli Pariser. *The filter bubble: What the internet is hiding from you.* Penguin UK, 2011.

[251] Michael Quinn Patton. *Qualitative research & evaluation methods: Integrating theory and practice.* SAGE Publications, 2014.

[252] Carlo Perrotta and Neil Selwyn. Automation, AI, and education: Why sociotechnical studies are urgently needed in educational technology research. *British Journal of Educational Technology*, 51(3):997–1004, 2020.

[253] Persado. Persado's AI platform for marketing language optimization. https://www.persado.com/, 2019. Accessed: 2021-08-15.

[254] Christy Pettey. How AI will transform project management. *Gartner*, 2019.

[255] Rosalind W Picard. *Affective computing*. MIT Press, 1997.

[256] Sarah Pink. *Doing sensory ethnography*. SAGE, 2015.

[257] Robert E Ployhart and Robert J Vandenberg. Longitudinal research: The theory, design, and analysis of change. *Journal of management*, 36(1):94–120, 2010.

[258] Athanasios Polyportis. A longitudinal study on artificial intelligence adoption: Understanding the drivers of chatgpt usage behavior change in higher education. *Frontiers in Artificial Intelligence*, 6:1324398, 2023.

[259] Veljko Potkonjak, Michael Gardner, Victor Callaghan, Pasi Mattila, Christian Guetl, Vladimir M Petrović, and Kosta Jovanović. Virtual laboratories for education in science, technology, and engineering: A review. *Computers & Education*, 95:309–327, 2016.

[260] Jenny Preece. *Online communities: Designing usability and supporting sociability*. John Wiley & Sons, Inc., 2000.

[261] John Preskill. Quantum computing in the NISQ era and beyond. *Quantum*, 2:79, 2018.

[262] Associated Press. AP's use of AI in journalism. https://www.ap.org/discover/artificial-intelligence, 2019. Accessed: 2021-08-15.

[263] James O Prochaska and Carlo C DiClemente. Stages and processes of self-change of smoking: Toward an integrative model of change. *Journal of Consulting and Clinical Psychology*, 51(3):390, 1983.

[264] UN Global Pulse. Un global pulse: Harnessing big data for development and humanitarian action. https://www.unglobalpulse.org/, 2019. Accessed: 2021-08-15.

[265] Junaid Qadir, Anwaar Ali, Raihan ur Rasool, Andrej Zwitter, Arjuna Sathiaseelan, and Jon Crowcroft. Crisis analytics: Big data-driven crisis response. *Journal of International Humanitarian Action*, 1(1):1–21, 2016.

[266] Charles C Ragin. *Redesigning social inquiry: Fuzzy sets and beyond*. University of Chicago Press, Chicago, 2008.

[267] Iyad Rahwan, Manuel Cebrian, Nick Obradovich, Josh Bongard, Jean-François Bonnefon, Cynthia Breazeal, Jacob W Crandall, Nicholas A Christakis, Iain D Couzin, Matthew O Jackson, et al. Machine behaviour. *Nature*, 568(7753):477–486, 2019.

[268] Salman Razzaki, Adam Baker, Yura Perov, Katherine Middleton, Janie Baxter, Daniel Mullarkey, Davinder Sangar, Michael Taliercio, Mobasher Butt, Azeem Majeed, et al. A comparative study of artificial intelligence and human doctors for the purpose of triage and diagnosis. *arXiv preprint arXiv:1806.10698*, 2018.

[269] Johnmarshall Reeve. Self-determination theory applied to educational settings. *Handbook of self-determination research*, 2:183–204, 2002.

[270] Justin Reich. *Failure to disrupt: Why technology alone can't transform education*. Harvard University Press, 2020.

[271] Kurt M Ribisl, Maureen A Walton, Carol T Mowbray, Douglas A Luke, William S Davidson II, and Bonnie J Bootsmiller. The effects of nonresponse on prevalence estimates for a referent population: A simulation study. *Journal of Community Psychology*, 24(2):95–113, 1996.

[272] Catherine Kohler Riessman. *Narrative methods for the human sciences*. SAGE, 2008.

[273] Lionel Robert. Personality in the human robot interaction literature: A review and brief critique. *Proceedings of the 24th Americas Conference on Information Systems*, 2018.

[274] Laura Robinson, Shelia R Cotten, Hiroshi Ono, Anabel Quan-Haase, Gustavo Mesch, Wenhong Chen, Jeremy Schulz, Timothy M Hale, and Michael J Stern. Digital inequalities and why they matter. *Information, Communication & Society*, 18(5):569–582, 2015.

[275] RoboKind. Milo: A robot for children with autism. https://www.robokind.com/robots4autism, 2019. Accessed: 2021-08-15.

[276] David Rolnick, Priya L Donti, Lynn H Kaack, Kelly Kochanski, Alexandre Lacoste, Kris Sankaran, Andrew Slavin Ross, Nikola Milojevic-Dupont, Natasha Jaques, Anna Waldman-Brown, et al. Tackling climate change with machine learning. *arXiv preprint arXiv:1906.05433*, 2019.

[277] Gillian Rose. *Visual methodologies: An introduction to researching with visual materials*. SAGE, 2016.

[278] Alexander J Rothman. Motivating health behavior change and maintenance: The self-regulation imperative. *Health Psychology*, 25(3S):S54, 2006.

[279] Cynthia Rudin. Stop explaining black box machine learning models for high stakes decisions and use interpretable models instead. *Nature Machine Intelligence*, 1(5):206–215, 2019.

[280] Stuart Russell. *Human compatible: Artificial intelligence and the problem of control*. Viking, 2019.

[281] Richard M Ryan and Edward L Deci. *Self-determination theory and the facilitation of intrinsic motivation, social development, and well-being*. American Psychologist, 2000.

[282] Safetipin. Safetipin: Making cities safer. `https://safetipin.com/`, 2020. Accessed: 2021-08-15.

[283] Eduardo Salas, Nancy J Cooke, and Michael A Rosen. On teams, teamwork, and team performance: Discoveries and developments. *Human factors*, 50(3):540–547, 2008.

[284] Nithya Sambasivan, Shivani Kapania, Hannah Highfill, Diana Akrong, Praveen Paritosh, and Lora M Aroyo. Re-imagining algorithmic fairness in india and beyond. *Proceedings of the 2021 ACM Conference on Fairness, Accountability, and Transparency*, pages 315–328, 2021.

[285] Ben D Sawyer, Victor S Finomore, Andres A Calvo, and Peter A Hancock. Google glass: A driver distraction cause or cure? *Human factors*, 56(7):1307–1321, 2014.

[286] Astrid Schepman and Paul Rodway. The measurement of attitudes towards artificial intelligence: An overview and recommendations. *The Impact of Artificial Intelligence on Societies: Understanding Attitude Formation Towards AI*, pages 9–24, 2024.

[287] Gregory Schraw and Rayne Sperling Dennison. Assessing metacognitive awareness. *Contemporary educational psychology*, 19(4):460–475, 2006.

[288] Dale H Schunk and Barry J Zimmerman. Self-regulation and learning. *Handbook of Psychology, Second Edition*, 7, 2012.

[289] ScriptBook. Scriptbook: AI-powered script analysis tool. `https://www.scriptbook.io/`, 2018. Accessed: 2021-08-15.

[290] Isabella Seeber, Eva Bittner, Robert O Briggs, Triparna de Vreede, Gert-Jan De Vreede, Aaron Elkins, Ronald Maier, Alexander B Merz, Sarah Oeste-ReiSS, Nils Randrup, et al. Machines as teammates: A research agenda

on AI in team collaboration. *Information & management*, 57(2):103174, 2020.

[291] Irving Seidman. *Interviewing as qualitative research: A guide for researchers in education and the social sciences.* Teachers College Press, 2013.

[292] Amartya Sen. Well-being, agency and freedom: The Dewey lectures 1984. *The Journal of Philosophy*, 82(4):169–221, 1985.

[293] Amartya Sen. *Development as freedom.* Oxford University Press, 1999.

[294] Burr Settles and Brendan Meeder. A trainable spaced repetition model for language learning. *Proceedings of the 54th Annual Meeting of the Association for Computational Linguistics (volume 1: long papers)*, pages 1848–1858, 2016.

[295] William R Shadish, Thomas D Cook, and Donald Thomas Campbell. *Experimental and quasi-experimental designs for generalized causal inference.* Houghton Mifflin Boston, 2002.

[296] Mark D Shermis. Contrasting state-of-the-art automated scoring of essays: Analysis. *Educational Measurement: Issues and Practice*, 32(2):3–13, 2013.

[297] Ben Shneiderman. Direct manipulation: A step beyond programming languages. *Computer*, 16(8):57–69, 1983.

[298] Ben Shneiderman. Human-centered artificial intelligence: Reliable, safe & trustworthy. *International Journal of Human–Computer Interaction*, 36(6):495–504, 2020.

[299] Siemens. Siemens uses artificial intelligence to reduce energy consumption. https://press.siemens.com/global/en/pressrelease/siemens-uses-artificial-intelligence-reduce-energy-consumption, 2017. Accessed: 2021-08-15.

[300] George Siemens and Phil Long. Learning analytics: The emergence of a discipline. *American Behavioral Scientist*, 57(10):1380–1400, 2013.

[301] SkillsFuture Singapore. Skillsfuture. `https://www.skillsfuture.gov.sg/`, 2019. Accessed: 2021-08-15.

[302] Judith D Singer and John B Willett. *Applied longitudinal data analysis: Modeling change and event occurrence*. Oxford University Press, 2003.

[303] Aaron Smith. Public attitudes toward computer algorithms. *Pew Research Center*, 16, 2018.

[304] SP Somashekhar, MJ Sepúlveda, S Puglielli, AD Norden, EH Shortliffe, C Rohit Kumar, A Rauthan, N Arun Kumar, P Patil, K Rhee, et al. Watson for oncology and breast cancer treatment recommendations: agreement with an expert multidisciplinary tumor board. *Annals of Oncology*, 29(2):418–423, 2018.

[305] Betsy Sparrow, Jenny Liu, and Daniel M Wegner. Google effects on memory: Cognitive consequences of having information at our fingertips. *Science*, 333(6043):776–778, 2011.

[306] Felicia Stokes and Amitabha Palmer. Artificial intelligence and robotics in nursing: Ethics of caring as a guide to dividing tasks between AI and humans. *Nursing Philosophy*, 21(4):e12306, 2020.

[307] Jonathan M Stokes, Kevin Yang, Kyle Swanson, Wengong Jin, Andres Cubillos-Ruiz, Nina M Donghia, Craig R MacNair, Shawn French, Lindsey A Carfrae, Zohar Bloom-Ackermann, et al. A deep learning approach to antibiotic discovery. *Cell*, 180(4):688–702, 2020.

[308] Stratejos. Stratejos: AI-powered project management assistant. `https://stratejos.ai/`, 2018. Accessed: 2021-08-15.

[309] Emma Strubell, Ananya Ganesh, and Andrew McCallum. Energy and policy considerations for deep learning in NLP. *arXiv preprint arXiv:1906.02243*, 2019.

[310] Lucy A Suchman. *Human-machine reconfigurations: Plans and situated actions.* Cambridge University Press, 2nd edition, 2007.

[311] Arun Sundararajan. *The sharing economy: The end of employment and the rise of crowd-based capitalism.* MIT Press, 2016.

[312] Daniel Susser, Beate Roessler, and Helen Nissenbaum. Technology, autonomy, and manipulation. *Internet Policy Review*, 8(2), 2019.

[313] John Sweller. Cognitive load during problem solving: Effects on learning. *Cognitive Science*, 12(2):257–285, 1988.

[314] Damian A Tamburri. The AI black box explicability conundrum: an empirical study. *Empirical Software Engineering*, 26(5):1–40, 2021.

[315] Toon Taris. *A primer in longitudinal data analysis.* SAGE, 2000.

[316] Abbas Tashakkori and Charles Teddlie. *SAGE handbook of mixed methods in social & behavioral research.* SAGE Publications, Inc., 2010.

[317] Charles Taylor. *The ethics of authenticity.* Harvard University Press, 1991.

[318] Century Tech. Century tech: AI-powered learning platform. https://www.century.tech/, 2019. Accessed: 2021-08-15.

[319] AIVA Technologies. AIVA: Artificial intelligence virtual artist. https://www.aiva.ai/, 2019. Accessed: 2021-08-15.

[320] Max Tegmark. *Life 3.0: Being human in the age of artificial intelligence.* Knopf, 2017.

[321] Textio. Textio: Augmented writing platform. `https://textio.com/`, 2018. Accessed: 2021-08-15.

[322] Richard H Thaler and Cass R Sunstein. *Nudge: Improving decisions about health, wealth, and happiness.* Yale University Press, 2008.

[323] Nenad Tomašev, Julien Cornebise, Frank Hutter, Shakir Mohamed, Angela Picciariello, Bec Connelly, Danielle CM Belgrave, Daphne Ezer, Fanny Camille van der Haert, Frank Mugisha, et al. AI for social good: unlocking the opportunity for positive impact. *Nature Communications*, 11(1):1–6, 2020.

[324] Eric J Topol. High-performance medicine: The convergence of human and artificial intelligence. *Nature Medicine*, 25(1):44–56, 2019.

[325] Ehsan Toreini, Mhairi Aitken, Kovila Coopamootoo, Karen Elliott, Carlos Gonzalez Zelaya, and Aad van Moorsel. The relationship between trust in AI and trustworthy machine learning technologies. *Proceedings of the 2020 Conference on Fairness, Accountability, and Transparency*, pages 272–283, 2020.

[326] Manuel Trajtenberg. Ai as the next GPT: A political-economy perspective. *National Bureau of Economic Research*, 2018.

[327] Shari Trewin. AI fairness for people with disabilities: Point of view. *arXiv preprint arXiv:1811.10670*, 2019.

[328] Eric Trist. The evolution of socio-technical systems. *Occasional Paper*, 2, 1981.

[329] Yaacov Trope and Nira Liberman. Temporal construal. *Psychological Review*, 110(3):403, 2003.

[330] Ilkka Tuomi. The impact of artificial intelligence on learning, teaching, and education. *Policies for the Future*, 1, 2018.

[331] Unilever. Unilever's flex experiences platform. `https://www.unilever.com/news/press-releases/2019/unilever-launches-new-ai-powered-talent-marketplace/`, 2019. Accessed: 2021-08-15.

[332] International Telecommunication Union. Measuring digital development: Facts and figures 2020. `https://www.itu.int/en/ITU-D/Statistics/Pages/facts/default.aspx`, 2020. Accessed: 2021-08-15.

[333] International Telecommunication Union. AI for good. `https://aiforgood.itu.int/`, 2021. Accessed: 2021-08-15.

[334] Alex Urmeneta and Margarida Romero. *Creative applications of artificial intelligence in education*. Springer Nature, 2024.

[335] Jan Van Dijk and Kenneth Hacker. The digital divide as a complex and dynamic phenomenon. *The Information Society*, 19(4):315–326, 2003.

[336] Kurt VanLehn. The behavior of tutoring systems. *International Journal of Artificial Intelligence in Education*, 16(3):227–265, 2006.

[337] Kurt VanLehn. The relative effectiveness of human tutoring, intelligent tutoring systems, and other tutoring systems. *Educational Psychologist*, 46(4):197–221, 2011.

[338] Francisco J Varela, Evan Thompson, and Eleanor Rosch. *The embodied mind: Cognitive science and human experience*. MIT Press, 1991.

[339] Peter-Paul Verbeek. *What things do: Philosophical reflections on technology, agency, and design*. Penn State Press, 2005.

[340] Stefaan G Verhulst. Reimagining democratic governance for the 21st century: The potential of AI. *AI & SOCIETY*, 35(3):745–750, 2020.

[341] Roumen Vesselinov and John Grego. *Duolingo effectiveness study.* City University of New York, USA, 28:1–25, 2012.

[342] Ricardo Vinuesa, Hossein Azizpour, Iolanda Leite, Madeline Balaam, Virginia Dignum, Sami Domisch, Anna Felländer, Simone Daniela Langhans, Max Tegmark, and Francesco Fuso Nerini. The role of artificial intelligence in achieving the sustainable development goals. *Nature communications*, 11(1):1–10, 2020.

[343] Susanne Vogl. Mixed methods longitudinal research. *Forum: Qualitative Social Research*, 24(1):Art. 21, 2023.

[344] Paul Voigt and Nils Hullen. How is the AI act implemented and enforced? In *The EU AI Act: Answers to frequently asked questions*, pages 151–181. Springer, 2024.

[345] Paul Voigt and Axel Von dem Bussche. The EU general data protection regulation (GDPR). A Practical Guide, 1st Ed. Cham: Springer International Publishing, 2017.

[346] Lev S Vygotsky. *Mind in society: The development of higher psychological processes.* Harvard University Press, 1978.

[347] Bernhard Wahl, Annette Cossy-Gantner, Stefan Germann, and Nina R Schwalbe. Artificial intelligence in medicine and cardiac imaging: Harnessing big data and advanced computing to provide personalized medical diagnosis and treatment. *Current Cardiology Reports*, 20(6):1–8, 2018.

[348] Douglas R Wassenaar. Ethical issues in social science research. *Research in Practice: Applied Methods for the Social Sciences*, 2:60–79, 2007.

[349] David Wechsler. *Wechsler adult intelligence scale–Fourth Edition (WAIS–IV)*. San Antonio, TX: NCS Pearson, 2008.

[350] Sarah Myers West, Meredith Whittaker, and Kate Crawford. Inclusive AI: Technology and policy for a diverse urban future. *AI Now Institute*, 2018.

[351] Norbert Wiener. *The human use of human beings: Cybernetics and society.* Houghton Mifflin, 1950.

[352] Rainer Winkler and Matthias Söllner. Alexa, are you my mom? a study on children's interactions with chatbots and the implications for conversational interface design. *Proceedings of the 38th International Conference on Information Systems (ICIS)*, 2017.

[353] Jochen Wirtz, Paul G Patterson, Werner H Kunz, Thorsten Gruber, Vinh Nhat Lu, Stefanie Paluch, and Antje Martins. Frontline employees' acceptance of and resistance to service robots in hospitality service encounters: The role of human–robot interaction. *Journal of Service Management*, 29(2):242–264, 2018.

[354] Jonathan R Wolpaw and Elizabeth Winter Wolpaw. Brain-computer interfaces for communication and control. *Handbook of Clinical Neurology*, 168:67–85, 2020.

[355] Wendy Wood and David T Neal. A new look at habits and the habit-goal interface. *Psychological Review*, 114(4):843, 2007.

[356] Brian Woods, David Adamson, Shayne Miel, and Elijah Mayfield. Formative writing assessment using automated essay scoring technology. *Journal of Writing Assessment*, 10(1), 2017.

[357] Beverly Park Woolf, H Chad Lane, Vinay K Chaudhri, and Janet L Kolodner. AI grand challenges for education. *AI Magazine*, 34(4):66–84, 2013.

[358] Samuel C Woolley and Philip N Howard. *Computational propaganda: Political parties, politicians, and political manipulation on social media*. Oxford University Press, 2018.

[359] Tessa Wright. The gendered impacts of technological change for public transport workers in the global south. *Research in Transportation Business & Management*, 31:100384, 2019.

[360] Hsin-Kai Wu, Silvia Wen-Yu Lee, Hsin-Yi Chang, and Jyh-Chong Liang. Current status, opportunities and challenges of augmented reality in education. *Computers & Education*, 62:41–49, 2013.

[361] Jack Yang, Mauricio Marrone, and Alireza Amrollahi. Actualising artificial intelligence affordances: A longitudinal study of a professional service firm. In *Proceedings of the 34th Australasian Conference on Information Systems*, pages 1–11, Wellington, New Zealand, 2023. AIS Electronic Library (AISeL).

[362] David Scott Yeager and Carol S Dweck. Mindsets that promote resilience: When students believe that personal characteristics can be developed. *Educational Psychologist*, 47(4):302–314, 2012.

[363] Robert K. Yin. *Case Study Research and Applications: Design and Methods*. SAGE Publications, 2018.

[364] Eliezer Yudkowsky. AI alignment: Why it's hard, and where to start. In Stuart Armstrong, editor, *Smarter than us: The rise of machine intelligence*, pages 1–14. Machine Intelligence Research Institute, 2016.

[365] Bluma Zeigarnik. On finished and unfinished tasks. *A source book of Gestalt psychology*, 1:300–314, 1938.

[366] Lingyu Zhang, Zhengran Ji, and Boyuan Chen. Crew: Facilitating human-ai teaming research. *arXiv preprint arXiv:2408.00170*, 2024.

[367] Zhi Zhou, Xu Chen, En Li, Liekang Zeng, Ke Luo, and Junshan Zhang. Edge intelligence: Paving the last mile of artificial intelligence with edge computing. *Proceedings of the IEEE*, 107(8):1738–1762, 2019.

[368] Marc A. Zimmerman and Julian Rappaport. Citizen participation, perceived control, and psychological empowerment. *American Journal of Community Psychology*, 16(5):725–750, 1988.

[369] Shoshana Zuboff. *The Age of Surveillance Capitalism: The Fight for a Human Future at the New Frontier of Power*. Profile Books, 2019.

Index

Note: Page references with *Italics* refer to figures.